CULTURAL ENCYCLOPEDIA OF THE BODY

Advisory Board

Carolyn de la Peña
University of California, Davis

Eugenia Paulicelli
Queens College, City University of New York

CULTURAL ENCYCLOPEDIA OF THE BODY

Volume 2: M–Z

Edited by
Victoria Pitts-Taylor

Kishwaukee College Library
21193 Malta Road
Malta, IL 60150-9699

GREENWOOD PRESS
Westport, Connecticut • London

Library of Congress Cataloging-in-Publication Data

Cultural encyclopedia of the body / edited by Victoria Pitts-Taylor.
  p. cm.
  Includes bibliographical references and index.
  ISBN: 978-0-313-34145-8 ((set) : alk. paper)
  ISBN: 978-0-313-34146-5 ((vol. 1) : alk. paper)
  ISBN: 978-0-313-34147-2 ((vol. 2) : alk. paper)

  1. Body, Human—Social aspects—History. 2. Body, Human—Social aspects—Dictionaries. 3. Body image—Social aspects. I. Pitts-Taylor, Victoria.
  HM636.C85 2008
  306.4—dc22   2008019926

British Library Cataloguing in Publication Data is available.

Copyright © 2008 by Victoria Pitts-Taylor

All rights reserved. No portion of this book may be reproduced, by any process or technique, without the express written consent of the publisher.

Library of Congress Catalog Card Number: 2008019926
ISBN: 978-0-313-34145-8 (Set)
    978-0-313-34146-5 (Vol. 1)
    978-0-313-34147-2 (Vol. 2)

First published in 2008

Greenwood Press, 88 Post Road West, Westport, CT 06881
An imprint of Greenwood Publishing Group, Inc.
www.greenwood.com

Printed in the United States of America

The paper used in this book complies with the Permanent Paper Standard issued by the National Information Standards Organization (Z39.48-1984).

10 9 8 7 6 5 4 3 2 1

For Chloe

CONTENTS

Volume 1

Preface	xiii
Acknowledgments	xv
Introduction	xvii
Chronology	xxix
The Encyclopedia	

Abdomen — 1
Abdominoplasty — 1

Blood — 4
Bloodletting — 4
Cultural History of Blood — 6

Brain — 12
Cultural History of the Brain — 12

Breasts — 21
Breast Cancer Awareness — 21
Breastfeeding — 25
Breast Ironing — 29
Cultural History of the Breast — 31
History of the Brassiere — 44
Lolo Ferrari's Breasts — 50
Surgical Reduction and Enlargement of Breasts — 52

Buttocks — 59
Cultural History of the Buttocks — 59
Surgical Reshaping of the Buttocks — 66

Cheeks — 69
Surgical Reshaping of the Cheekbones — 69

Chest 72
O-Kee-Pa Suspension 72
Pectoral Implants 74

Chin 77
Surgical Reshaping of the Chin 77

Clitoris 80
Cultural History of the Clitoris 80

Ear 86
Cultural History of the Ear 86
Ear Cropping and Shaping 91
Ear Piercing 93
Earlobe Stretching 100
Otoplasty 102

Eyes 105
Blepharoplasty 105
Cultural History of the Eyes 110
History of Eye Makeup 116
History of Eyewear 122

Face 127
Cultural Ideals of Facial Beauty 127
History of Antiaging Treatments 137
History of Makeup Wearing 141
Medical Antiaging Procedures 148
Michael Jackson's Face 159
Performance Art of the Face: Orlan 161

Fat 165
Cellulite Reduction 165
Cultural History of Fat 167
Dieting Practices 172
Eating Disorders 178
Fat Activism 183
Fat Injections 189
Surgical Fat Reduction and Weight-Loss Surgeries 191

Feet 197
Feet: Cultural History of the Feet 197
Footbinding 201
History of Shoes 208

Genitals 216
Cultural History of Intersexuality 216
Female Genital Cutting 224
Genital Piercing 231
Sex Reassignment Surgery 235

Hair — 246
Cultural History of Hair — 246
Hair Removal — 256
Hair Straightening — 260
Shaving and Waxing of Pubic Hair — 263

Hands — 266
Cultural History of the Hands — 266
Mehndi — 274
The Hand in Marriage — 276

Head — 281
Cultural History of the Head — 281
Head Shaping — 288
Head Slashing — 290
Veiling — 292

Heart — 298
Cultural History of the Heart — 298

Hymen — 306
Surgical Restoration of the Hymen — 306

Jaw — 308
Surgical Reshaping of the Jaw — 308

Labia — 311
Labiaplasty — 311

Legs — 314
Cultural History of Legs — 314

Limbs — 321
Limb-Lengthening Surgery — 321

Lips — 325
Cultural History of Lips — 325
Lip Enlargement — 331
Lip Stretching — 333

Volume 2

Mouth — 337
Cultural History of the Mouth — 337

Muscles — 344
Cultural History of Bodybuilding — 344

Neck — 351
Neck Lift — 351
Neck Rings — 353

Nipple — 357
Nipple Removal — 357

Nose — 360
Cultural History of the Nose — 360
Nose Piercing — 366
Rhinoplasty — 370

Ovaries — 379
Cultural History of the Ovaries — 379

Penis — 384
Cultural History of the Penis — 384
Male Circumcision — 390
Penis Envy — 396
Subincision — 398

Reproductive System — 402
Cultural History of Menopause — 402
Cultural History of Menstruation — 407
Fertility Treatments — 411
History of Birth Control — 418
History of Childbirth in the United States — 428
Menstruation-Related Practices and Products — 434

Semen — 437
Cultural History of Semen — 437

Skin — 446
Body Piercing — 446
Branding — 452
Cultural History of Skin — 454
Cutting — 464
Scarification — 466
Skin Lightening — 473
Stretch Marks — 476
Subdermal Implants — 478
Tattoos — 479

Skull — 490
Trephination — 490

Teeth — 493
Cosmetic Dentistry — 493
Teeth Filing — 497

Testicles — 502
Castration — 502
Eunuchs in Antiquity — 507

Thigh — 512
Liposuction of the Thigh — 512

Tongue	514
Tongue Splitting	514
Uterus	517
Chastity Belt	517
Cultural History of the Vagina	519
The "Wandering Womb" and Hysteria	526
Vagina	532
The Vagina in Childbirth	532
The Vagina Monologues	536
Vaginal-Tightening Surgery	537
Waist	540
History of the Corset	540
Selected Bibliography	546
Index	561
About the Editor and Contributors	575

Mouth

Cultural History of the Mouth

The mouth is the point of entry into the body. As such it is not simply a physiological mechanism for eating or speaking; rather, it derives its many cultural meanings from the fact that what enters the mouth becomes a part of the person. As an important boundary to the body, there are many rich cultural ideas associated with the history of the mouth and teeth.

The symbolism of the mouth goes well beyond its simple physiological functions. The mouth can be seen to have five different meanings that can be detected in ancient and more recent popular cultures. The first association of the mouth is with a hero's passageway to death. In this respect, entry into various mouths is associated with death and the loss of one's self.

The association between the mouth and death is often seen in different cultural depictions. Throughout history, there are associations of the mouth with dismemberment; in addition, the mouth has formed part of an illustration of the gate to the underworld and the jaws of hell, as well as the terrible devouring mother. In the Sanskrit text the *Bhagavad Gita* (ca. 500 BCE), for example, Lord Krishna is said to have revealed his terrible celestial forms, one of which involving the image of many mouths crushing into powder the heads of mortals and worlds. Likewise, in the Epic of Gilgamesh, an ancient Mesopotamian poem, a great monster called Humbaba, whose jaws were equated with death, was said to have a mouth full of breath like fire. There are many other examples of the mouth as a devourer of people. For example, a shamanic amulet carved in walrus tusk, associated with the Tlingit Indians of Alaska, has depicted a terrible mother figure about to eat a person held in her hands. Aztec myths include the great serpent Quetzalcoatl, which holds a person in its jaws. Likewise, in the Christian tradition during the eleventh century CE, a drawing from Constantinople depicts a ladder rising to heaven; at the top of the ladder is the image of Christ, to whom people on the ladder are ascending. Some are falling off the ladder into the mouth of a dragon that is waiting below. Throughout the history of Christianity, there appear repeated images of the "hells mouth" where mortals are consumed. For example, in a wood engraving, ca. 1540, by Lucas Cranach the Younger, where the true Church of the reformation is depicted as listening to Martin Luther and where the pope, cardinals, and friars are engulfed in the fire of "Hells mouth."

While the mouth is associated with death, it is also associated with rebirth. For example, in a myth associated with the Manja and Banda of Africa, a hairy black being called Ngakola was said to live in the bush. Ngakola devoured people and would then vomit them back out in a transformed state. Such stories are closely related to the growth and passage of adolescents into adulthood, when an adolescent's identity was swallowed and eaten, transformed within the belly of the beast and disgorged to become full members of the adult community. In this respect, the mouth is not only associated with the destruction of the old person but also has the transforming potential to produce a new identity. This kind of story is common across Polynesian, Australian aboriginal communities, and tribal cultures in North America, South America, and Africa. The most obvious example of such a literary trope in Western cultures is the biblical story of Jonah. Jonah first appears in the Old Testament but is also cited in the New Testament and is treated as a prophet in the Koran. Sailing to a place called Tarshish, Jonah is thrown overboard his vessel by sailors who realize he is to blame for causing a storm to swell in the sea. He is swallowed by a great fish and is saved when God instructs the fish to vomit him. In these examples, the mouth is associated with the reborn hero, and indeed the story remains important in contemporary culture. The most celebrated example of the transformed hero entering a mouth is in the Francis Ford Coppola film *Apocalypse Now*, based on the Joseph Conrad novel *Heart of Darkness*, published in 1902. In both versions of the story, the hero Marlow travels up the mouth of a river into the belly of the beast, a jungle, where horrible events lead to his transformation and rebirth.

Throughout history, the mouth has been associated with the pathway to destruction and as a canal for rebirth. The mouth also has been associated with another set of meanings. In a Swahili creation story, said to be heavily influenced by Islam, God was said to have spoken a word that subsequently enters through Adam's mouth and spreads life throughout his body. In this respect, the mouth can be associated with creation and the giving of life. In Japanese and Chinese mythology, the core symbols of wholeness, the sun, moon, the pearl, among others, which are associated with the bringing to consciousness, were originally held in the mouth of a dragon but had been "wrestled" free by people. As a consequence, the great dragon continues to pursue the orb, a depiction that forms the basis of many popular themes in Chinese and Japanese art. Likewise, the Eye of Horus in Egyptian myths is composed of the image of the sun within a mouth, something that has been said to be associated with the creative word. The Ouroboros, a dragon or serpent-like figure with its tail in its mouth, is an ancient symbol that has appeared in many mythical traditions since antiquity. Plato described such a creature as immortal since it could feed on itself. The Greeks may have borrowed the Ouroboros, or "tail devourer," from ancient Egypt. Hindu tradition holds that Shesha is a many-headed serpent from whose muliple mouths Vishnu is praised. Finally, in Asian cultures, traditional healers are often said to breathe in deeply the souls of ancestors who will in turn come to aid them in the process of healing.

While the mouth can be understood as a pathway through which life and the soul enters the body, conversely it can also be understood as the canal through which the soul escapes. This depiction has been demonstrated in eleventh-century German art in which a woman was depicted as dying in her bed while her naked soul escaped from her mouth. Likewise, the soul of St. John was depicted as escaping from his mouth and into the presence of Christ in ninth- and tenth-century Italian art. The risk that the soul might flee the body has been said to accompany yawning in the Indian tradition, and it was though that one way to avoid this was to snap ones fingers. Finally, in many medieval depictions of possession, the mouth became the exit point of fleeing demons.

A final broad meaning associated with the mouth is that of complete consciousness. In the German legend of St. John Chrysostom, a youth who was subject to the ridicule of others but who prayed every morning was one day called to kiss the mouth of the Virgin. The call went out several times before he responded; upon kissing her mouth, he was filled with great wisdom and the ability to speak clearly. Additionally, a golden "circlet" formed around his mouth and was reported to have shone like a star. In the ancient Chinese text the *I Ching*, a hexagram called "I," otherwise known as "the corners of the mouth," is associated with a degree of wholeness, a nourishment of the self and others.

Anthropological Understandings of the Mouth

The mouth, in cultural anthropology, is generally considered to have significance because it represents a point of entry to and exit from the body. The mouth is therefore closely related to how the senses are organized through its association with taste and speech. It is the site of a bidirectional flow of both food into the body and speech out of the body. The degree to which this boundary is open has different consequences for the way in which a society is organized and indeed for how different groups respond to practices associated with the mouth. For example, many premodern cultures are often described as "oral;" in this respect, the boundary between what is inside the body and what is outside is permeable and uncontrollable. In such cultures, the mouth often features as an important cultural symbol. An extreme example of this was the oral culture of the Kwakiutl, whose world was filled with the mouths of all kinds of animals that would kill and destroy to satisfy their hunger. There were images of the gaping maws of killer whales and the snapping teeth of wolves and bears. There were many images of mouths greedy for food, women who would snatch unwary travelers and eat them, and children who, while quietly sucking on their mothers' breasts, would turn into monsters and devour them. Likewise, in contemporary culture, the permeability of the mouth has important consequences leading to different rules about what people will or will not do with their mouths in relation to things such as sharing a toothbrush with family members versus complete strangers to what they will put into their mouths during sexual intercourse.

Because of the importance of the bidirectional flow of eating and speech, the mouth has often been subjected to various forms of regulation, and as a result, different emphasis is often placed on the varying sides of the bidirectional flow. This has consequences for the role of the mouth in society. One culture can value ingestion and another speech. This cannot be more profoundly observed in relation to the way in which the meal is organized in premodern and modern societies.

Eating in societies is often structured into a shared form, the meal. In so-called premodern societies, eating was associated with being "eaten into the community" and the sharing of companionship. At the same time as the food was eaten, the person became incorporated into the community with little distinction between the person and the society in which they existed. Everyone was the same and shared the same identity. Often the outflow of speech in such cultures would receive less attention; for example, members of families in premodern times were often required to sit together while eating the same food, and often in silence. This indicates a greater emphasis on sharing and ingestion than on speech.

Anthropologists and sociologists alike argue that the change from premodern to modern cultures is primarily characterized by a "closing of the mouth," and that this has had important consequences for the way in which meals are organized. As described, in premodern cultures, everyone shared and took part in eating the same meal. In doing so, they celebrated being part of the same culture where everyone had the same identity. In modern cultures, however, the emphasis has shifted to focus on helping people become individuals. To this end, the family meal as a form of shared communion between similar people has become marginalized. Indeed, as a consequence, new ritual forms of eating have emerged, including the snack and other types of food that are not considered "real" food, including confectionary, sweets, and alcohol. Indeed, some forms of oral consumption can be considered to be largely empty of nutrition (tobacco and chewing gum). While the existence of confectionary can be traced back as far as Egyptian times, such consumption was more a marker of something shared in the form of a gift than as part of a ritual geared toward the pleasure of individuals. In this respect, changes associated with the closing of the mouth have had consequences for the social organization of eating. The emphasis on eating is no longer about the consumption of the same meal. For example, when people go out to a restaurant, each person chooses what *they* want to eat, and the emphasis is on their ability to share in good conversation. So while the mouth has become closed with respect to eating, it has become open with respect to sharing in conversation. Likewise, emphasis has shifted from the closeness of sharing the same food to the more distant sharing of the same conversation. In many ways, the transformation of the mouth from being open to closed also has marked itself onto the nature of the conversation, which in turn involved a great deal of self-control.

The Teeth

The most direct cultural associations with teeth are the themes of power, strength, and destruction. In images where the mouth was associated with death, the teeth often featured heavily. Other examples of such imagery include Greek mythology, when a Phonecian prince who is said to have founded the city of Thebes, Cadmus, acting on the advice of Pallas, sowed the earth with the teeth of a dragon. Subsequently, a race of fierce men emerged from the earth only to kill each other. The destructive association of teeth is popularized almost universally in various folklores through the concept of vampirism. Here, the teeth of the undead are implicated in the consumption of the living. Vampirism is referenced in many cultural traditions, including Eastern European Slavic cultures. Much of the published folklore on the subject dates to the eighteenth century, but vampires figured heavily in oral traditions throughout medieval Europe.

Teeth are associated with animalistic qualities, and such associations, when valued, are often enhanced through various practices such as filing the teeth. Many tribal cultures across the world have engaged in such pursuits, including in Africa, North America, Mesoamerica, South America, India, Southeast Asia, the Malay Archipelago, the Philippines, and Japan. Early Mayans (2000 BCE) sharpened their teeth, and later Mayans also sometimes inlaid their teeth with jade. The Ticuana tribe of Brazil files children's teeth to sharp points, which is considered a mark of great beauty. Other tribes of the Amazon basin file teeth in order to enhance their ferociousness. Various tribes in Africa, such as the Ibo and the Tiv from Nigeria, have filed their teeth for beautification. In many cases, teeth filing is an initiatory rite that marks the passage into adulthood. In various encounters between colonizers and native groups, teeth filing has been controversial and sometimes misunderstood as a sign of cannibalism. Colonial governments have occasionally moved to ban the practice, as in Indonesia.

In many cultures, teeth are considered a source of power and prowess. After all, teeth often survive long after other bodily matter has been dissolved. Likewise, in some cultures, teeth are often valued for their association with inner strength. The Yuga of the Amazon, a war-like tribe, often removed teeth from the decapitated heads of their enemies to make necklaces. These became closely guarded heirlooms handed down from generation to generation, in the hope that they would bring prosperity and power to the family. In contemporary cultures, data from ethnographic research has demonstrated important associations between teeth and youthfulness. Indeed, some images of aging included feelings of remorse, a loss of strength, and acute experiences of one's mortality that accompanies the loss of a tooth. The loss of a tooth has been called a "symbolic little death," to stand as a metaphor for impotence and a punishment for sexual guilt. Sigmund Freud (1856–1939) theorized that dreams depicting the loss of teeth were suggestive of fear of castration, perhaps as punishment for masturbation.

Teeth can also reflect a falseness of character. Such sentiments have a long history. For example, the Roman satirists Martial and Juvenal, writing in the first and second centuries CE, wrote that while Lucania had white teeth and Thaï brown, one had false teeth and the other their own. In this respect, the authenticity of the brown tooth was contagious toward the character of the person concerned; likewise, the false white tooth led to assumptions of a falseness of character. Today, however, white teeth are often considered an important aspect of North American culture, as in the example the "American Smile" lauded in *Time* magazine during the 1990s as a key symbol of

America's greatness. The American smile with its pristine, manufactured teeth was affectionately contrasted to the soft, broken, and crooked textures of the "British smile" in *The Simpsons* episode "Last Exit to Springfield." While the pristine, white smile might be valued, in some cultures, both contemporary and historic, the opposite has been the case. Recent research clearly demonstrates contrasting views of the importance of a white smile in British society: in one case, a white smile was associated with a shallow and fake personality. In his Gothic tales, the nineteenth-century Italian writer Iginio Ugo Tarchetti (1839–1869) reported on the contrast between the rebelliousness and ferocity of white teeth in comparison to the sweetness and affection he felt toward black teeth. Indeed, the image of horribly long, white teeth was exposed by a curved grimace ready to bite, seize, and lacerate living flesh. White teeth gave the face a savage look, he suggested, whereas black, stubby teeth suggested a meek and homely nature.

The ways in which the teeth can become important cultural markers of the differences between people can be a powerful indicator of social division and status in society. During the sixteenth to the eighteenth centuries, tooth decay was largely considered a disease of the rich. During this period, a power struggle between mercantile and consumer capitalism saw the evolution of sugar, the main protagonist of tooth decay in the modern world, from a rare spice to a mass-produced and consumed product. With this transformation, what was once a disease of the rich became a disease of poverty. This had important consequences for the meaning of tooth decay and its representations in art and literature.

Power, Pain, and Dentistry

Pain is one of the most enduring associations with teeth throughout the modern world, and it is socially patterned and experienced. Sixteenth-century European art almost exclusively depicts the sufferers of tooth decay as poor. Here, the association between the loss of teeth and the loss of power was often reflected the social position of the poor who were to suffer ritual humiliation and usually in public. Contemporary art in the Netherlands reflects many of these important themes. For example, a picture by Leonard Beck (1480–1542), called *The Dentist* (ca. 1521), shows the image of a dentist pulling the tooth of a poor woman in a public place. The dentist brandishes the tooth high in the air triumphantly as if the extraction was painless and easy. The traditional relationship of dentists and their public therefore was formed on the basis of deception. Other images from the same early modern time provide further illustration of this important theme. Particularly striking is the image produced by Hans Sebald Beham (1500–1550), provided in his *Detail of the Great Country Fair* (1539). In this image, a tooth is being extracted from a poor person at a public fair, and while the person is distracted, he is robbed. In this respect, early images associated with the extraction of teeth were associated with charlatanism. Indeed, an old proverb was said to reflect that someone could lie "like a tooth puller." It is no accident that almost universally in these early images it is the poor who are depicted being subjected to pain and suffering by a different class of people, the dentists.

The power division between dentists and their patients becomes crucial to the imagery of the relationship between classes in Europe during the seventeenth century. These depictions not only demonstrated differences between the wealthy and the poor but also between town and countryside. Despite these changes, the social division between dentists and their patients persisted and from time to time the relationship could become one of outright hostility. A good example of this is Thomas Rowlandson's (1756–1827) *Transplanting of Teeth* (1787), in which a dentist is depicted to be extracting the teeth of the poor in order to transplant them into the mouths of the rich. Over time, however, the image of the experience of tooth pulling changed. The experience of pain became universal and was applied to all sections of society. Eventually, tooth pulling retreated indoors and became a private affair. In many respects, it could be argued that the changes that led to a closing of the mouth led inevitably to a closing of doors and a shift from public humiliation to a private space.

The care of teeth by professional dentists has expanded in the twentieth and twenty-first centuries to include cosmetic enhancements, leading to debates about the role of dentistry in promoting health. Contemporary cosmetic dentistry includes transforming the alignment of teeth, changing the color of teeth (exclusively to white), and exchanging false teeth, through dentures or implants, from unsightly teeth. *See also* Teeth: Cosmetic Dentistry; and Teeth: Teeth Filing.

Further Reading: Benveniste, Daniel. "The archetypal image of the mouth and its relation to autism." *The Arts in Psychotherapy* 10 (1983): 99–112; Falk, Pasi. *The Consuming Body.* London: Sage, 1994; Gregory, Jane, Barry Gibson, and Peter Glen Robinson. "Variation and change in the meaning of oral health related quality of life: A 'grounded' systems approach." *Social Science and Medicine* 60 (2004): 1,859–1,868; Kunzle, David. "The Art of Pulling Teeth in the Seventeenth and Nineteenth Centuries: From Public Martyrdom to Private Nightmare and Political Struggle?" In *Fragments for a History of the Human Body: Part Three*, M. Feher, R. Naddaff, and N. Tazi, eds. Cambridge, MA: MIT Press, 1989; Nations, Marilyn K., and Sharmênia de Araújo Soares Nuto. "'Tooth worms,' poverty tattoos and dental care conflicts in Northeast Brazil." *Social Science & Medicine* 54 (2002): 229–244; Nettleton, Sarah. *Power, Pain and Dentistry.* Buckingham: Open University Press, 1992; Thorogood, N. "Mouthrules and the Construction of Sexual Identities." *Sexualities* 3 (2000): 165–182.

Barry Gibson

Muscles

Cultural History of Bodybuilding

Bodybuilding is a form of modifying the body by building muscle mass. Through specialized exercise and training, accompanied by a regulated diet, a person can reshape and control muscle distribution throughout his or her body. Bodybuilders often follow extreme diet and exercise regimens. Using exercise, bodybuilders train specific body parts during each workout, alternating areas of the body. Bodybuilding workouts are controlled to include a certain number of sets and repetitions specific to each exercise. The number of sets refers to how many times certain repetitions, or the same flexing movement, will be done. Bodybuilders often do not focus much on cardiovascular workouts, except to burn calories in order to decrease their body fat so that their muscles will appear more defined. Serious bodybuilders work out five to seven times per week and follow a rigorous diet, which usually consists of large amounts of protein. Bodybuilding can be done for purely aesthetic reasons or for sport, in which case it is called competitive bodybuilding. Bodybuilding is not wholly about strength; in fact, it is more a sport of aesthetics. In competitive bodybuilding, there is no weight lifting portion of the competition.

History

Although bodybuilding is practiced by both men and women, bodybuilding historically reflects dominant cultural ideas of masculinity and the male body. Bodybuilding can be traced to ancient Greece and Rome, where the male body was carefully cultivated. In fact, illustrations on marble and other hard stone of people lifting weights date back to 3400 BCE, in the temples of ancient Egypt. During the original Olympic games, which began during the seventh century BCE, in Olympia, Greece, and lasted until almost the fourth century CE, men competed naked in order for the audience to see all of the contours of their bodies. The games were not just for sport, but also for the celebration of the human body in honor of the Greek god Zeus. During the gladiator competitions, the first of which took place in Rome in 264 BCE, slaves and war captives fought against each other, joined on occasion

Eugen Sandow lifting the "human dumbell." Courtesy of Library of Congress, LC-USZC4-6075.

by lions and other large animals. During these competitions, audience members reveled in the athletes' bodies.

The representation of muscular male physiques, particularly in sculpture, reached its peak in ancient Greece. These art works depicted the idealized male body. As ancient Rome became more powerful, Roman artists focused on more realistic depictions and images, creating statues and busts of their leaders, as well as realistic depictions of everyday life.

With the fall of the Roman Empire and the rise of Christianity, the portrayal of the human body changed. Christian artists depicted the body as imperfect and flawed. Increasingly, these artists began to cover the human body—emphasizing the shame that Adam and Eve felt after being cast from the Garden of Eden. When interest in science, technology, and the abilities of man was renewed during the Renaissance, artists rediscovered the human body. Their work demonstrated the influence of the art of ancient Greece, combined with the symbolism of Christian artwork. Renaissance artists such as Michelangelo (1475–1564) and Leonardo da Vinci (1452–1519) were fascinated with the contours of the athletic male body. One of the most well-known works of art to have been created during this time was Michelangelo's *David*, based on the biblical tale of David and Goliath. Even today, *David* is considered by many to be the perfect athletic ideal of the male body.

Though the muscular male body was celebrated in Renaissance art, it was not seen as ideal during the Victorian era, when tanned, muscular bodies were often associated with the working classes, reflecting hard labor and outdoor work. Today, building one's musculature is often an activity of leisure and privilege. Now that the lower classes tend to have higher rates of obesity and diabetes, displaying a lithe body signifies that one is able to afford healthful foods and has leisure time to attend a gym or yoga classes. While muscular bodies are currently in vogue, however, bodybuilding is still sometimes thought to embrace an extreme physical aesthetic.

Eugen Sandow and the Early Years of Bodybuilding

Bodybuilding as it is known today, or the building of the muscles for athletic or aesthetic reasons as a sport, is commonly thought to have come about during the late nineteenth century, growing in popularity in the early twentieth century. Bodybuilding's modern history begins with the emergence of Eugen Sandow (1867–1925), known as the father of modern bodybuilding, who is credited with popularizing the idea of developing the body purely for aesthetic reasons. Born Friederich Wilhelm Mueller in Prussia, Sandow was hired by entertainment promoter Florenz Ziegfeld Jr. and traveled with a sideshow, performing feats of strength such as picking up large objects and bench pressing. Sandow quickly learned that people were more fascinated by his physique than his feats. Though Sandow was muscular, he lacked the extreme frame that many bodybuilders now have. Sandow was not bulky; rather, his body resembled that of the Greek and Roman statues of warriors, Olympians, and gladiators. These works of art fascinated Sandow. He described the Grecian ideal as the perfect man. Sandow measured the proportions of these statues and calculated what the perfect male body should look like. His ideal measurements of neck, hip, bicep, thigh, calf, and chest sizes were based on height and bone structure. Sandow traveled throughout Europe and the United States displaying his physique and modeling, all the while penning books on nutrition and exercise. In 1897, Sandow opened his gym, the Institute of Physical Culture in London, and he trained the employees. Sandow greatly influenced other physical-strength enthusiasts. In 1898, *Health and*

Strength Magazine was founded in England by Hopton Hadley, a British entrepreneur.

In 1901, Sandow organized the Great Competition, the first large-scale bodybuilding contest in London, where men displayed their physiques in front of a panel of judges and an audience of over 15,000. During this competition, the proportion of the body was judged along with the size of the muscles. Men also displayed feats of strength by weight lifting as well as athletic prowess through their wrestling and fencing capabilities. Other small-scale competitions followed all over Great Britain and the United States. Sandow spent the rest of his life traveling and displaying his strength to audiences around the world and promoting his passion for strength and fitness. After his death from complications of syphilis, organized bodybuilding faded away, not to return until after the Great Depression and World War II.

Health and Strength Magazine organized the first Mr. Universe competition in 1948. Winners have included actor Steve Reeves and seven-time winner Arnold Schwarzenegger (1947–). In 1940, the first Mr. American competition was sponsored by the Amateur Athletic Union (AAU). In 1946, the International Federation of Bodybuilding (IFBB) was founded by Joe and Ben Weider. The IFBB sponsored the first Mr. Olympia competition for professional bodybuilders in 1965. The bronze statue given to winners is modeled after Sandow.

Beginning with the competitions in the late 1940s, the aesthetic of bodybuilding shifted from Sandow's image of the "perfect," idealized body. Men became larger and stronger, more muscles were expected, and bodybuilding became more of a highly competitive sport. In 1970, Sergio Oliva beat returning champion Arnold Schwarzenegger with what became the new ideal, an extremely muscular, V-shaped frame, with a large chest and narrow waist. Currently, exceptionally large and well-defined muscles are required in bodybuilding competitions. Competitions are based on the sheer size of the competitors, not the strength or their level of fitness. In the mid-1990s, however, competitions were organized that increased the focus on physical fitness. Bodybuilding as a sport and hobby is now global, with professional and amateur competitions in Asia, Africa, South America, and other Western nations.

Female Bodybuilding

Female bodybuilding has been controversial since its inception. Though women are allowed into competitions and judged on their size and the definition of their muscles, female bodybuilders have been held to a different standard. For much of the history of female bodybuilding, competitors were judged not only on their muscles, but also on their acceptability as conventionally feminine and attractive.

After World War II, some women began to take up bodybuilding. One was Abbye Stockton. She was known for flexing her muscles on the beaches of southern California and became very influential in developing female bodybuilding as a legitimate sport. In 1965, the National Amateur Bodybuilders Association (NABBA) organized the Miss Universe competition for female bodybuilding amateurs. In 1979, George Snyder organized a contest entitled

"Best in the World." Early female bodybuilding competitions, however, were little more than beauty pageants. In Snyder's competition, for example, women were instructed to wear high heels, makeup, and parade around the stage, and were prohibited from flexing their muscles. An early contestant, Carla Dunlap-Kaan, rebelled, and flexed her muscles in the same way her male counterparts did in their competitions. This brought Dunlap-Kaan and the sport of women's bodybuilding a lot of attention, and eventually paved the way for other female bodybuilders.

In these early events, women wore high heels and makeup, a tradition that continues in contemporary contests. The contestants were often thin, and at most, toned. Their bodies lacked the large muscles required in male competitions. However, during the "Golden Age of Bodybuilding," which began in the 1970s, women began to increase in size. During the 1983 Caeser's Cup competition, Bev Francis, a very large and muscular woman and a former power lifter, worked to have a similar size to that of male bodybuilders. However, she was declared not feminine enough and lost.

Amateur and professional female bodybuilding took off in the 1980s. Currently, there are female bodybuilding competitions all over the world. During the late 1980s and into the 1990s, female bodybuilders became a little more muscular, but feminine looks remained important in competitions. It was not until the 1990s that women were judged for their size instead of solely on their beauty. In 1995, the Fitness Olympia competition for women was organized by Debbie Kruck, which emphasized physical fitness as well as size.

Female bodybuilding is a controversial sport largely because it violates gendered beauty ideals. Female bodybuilders are expected to posses a very masculine body type in order to compete, while also are pressured to maintain a stereotypically feminine physique. Female bodybuilders can see their breasts flatten, shoulders become broad, and their necks become thick. Some cease to menstruate because of their low percentage of body fat, and experience an increase in testosterone, which sometimes gives them more masculine features. Because it violates gendered norms, the sexual orientation of female bodybuilders is sometimes questioned, and sexism and homophobia have been wielded against female bodybuilders. Some feminist writers have celebrated female bodybuilding, however, contending that through bodybuilding, women strengthen and empower themselves and rebel against male oppression and standard ideals of feminine beauty. These writers credit female bodybuilders with creating their own, postmodern standards of beauty.

Steroids

Anabolic steroids, a type of steroid hormones, were first identified and synthesized in Germany in the 1930s. Human anabolic steroids include Anavar, Durabolin, testosterone, Anadrol, Deca, Halotestin, Sustanon, and Andriol. Steroids were originally used to help people who were chronically sick or who had diseases that decreased muscle mass. Steroids help to greatly increase muscle mass as well as hair growth. Steroids also decrease female hormones in both males and females.

A number of bodybuilders and athletes take steroids in order to enhance their ability in their sport. Bodybuilders sometimes take steroids in order to quickly increase bulk and muscle mass, as well as to gain definition. Bodybuilding steroids are illegal and in the United States are known as Class III level drugs. People generally obtain steroids on the street or through other underground means. Doctors are prohibited by law from prescribing steroids solely for the purposes of bodybuilding and muscle enhancement. The steroids designed for animals but illegally used by humans are known as veterinary anabolic steroids, and these include Winstrol-V, Desposterona, Norandren 200, Equipoise, and Laurabolin. Both veterinary and human anabolic steroids have similar effects on the body; however, since veterinary steroids are less regulated, they can be more dangerous.

Bodybuilding in Popular Culture

Bodybuilding was highly publicized in the 1977 film *Pumping Iron*, which documented the 1975 Mr. Olympia bodybuilding competition. This film starred competitors Arnold Schwarzenegger, Franco Columbu, and Lou Ferrigno (who played the Hulk in the 1970s television series *The Incredible Hulk*). The documentary examined how they prepared for the competition, showing them practicing and competing. *Pumping Iron* was filmed during the "Golden Age" of bodybuilding, a time when bodybuilding and its athletes were gaining in popularity. During this era, the body builders were not as focused on bulk as much as they are in today's competitions, but rather on physique and proportion.

The sequel film, *Pumping Iron II: The Women*, was released in 1985 and featured female bodybuilders. This film examined the women who trained to compete in the Caeser's World Cup in 1983, a competition in which both professionals and amateurs competed. Notable bodybuilders such as Bev Francis, Rachel McLish, Lydia Cheng, and Lori Bowen were featured in this film. The bodybuilding competition in *Pumping Iron II* included women who were significantly smaller than the women who participate in contemporary competitions.

Many other visual representations of bodybuilding have been portrayed on television and in the movies, with the muscular Sylvester Stallone playing the title characters in *Rocky* and its numerous sequels, as well as *Rambo*, and its sequels. On television, the series *The Incredible Hulk*, starring Lou Ferrigno, was a hit as was the 1980s show *The A-Team*, featuring bodybuilding enthusiast Mr. T.

Body Image

Some researchers believe that some men who participate in bodybuilding activities suffer from muscle dysmorphia. While most body-image research has focused on women and examined anorexia and bulimia, recent research has examined body dysmorphia among men, including muscle dysmorphia or so-called bigorexia. Muscle dysmorphia is a form of obsessive-compulsive disorder in which people obsess over the size of their muscles. Some psychologists and psychiatrists claim that muscle dysmorphia is both underresearched and underdiagnosed.

The rise in interest in male bodybuilding may be linked to muscle dysmorphia, or it may be a cultural phenomenon. Feminist scholar Susan Bordo argues that men are increasingly under the kind of pressure to achieve the ideal body that women have been enduring for decades. Today, men are now being bombarded with advertisements, television, magazines, and other media with depictions of the so-called perfect male body.

Further Reading: Bordo, Susan. *The Male Body: A Look at Men in Public and in Private*. New York: Farrar, Straus and Giroux, 1997; Chapman, David L. *Sandow the Magnificent: Eugen Sandow and the Beginnings of Bodybuilding*. Champaign: University of Illinois Press, 1994; Moore, Pamela L., ed. *Building Bodies*. New Brunswick, NJ: Rutgers University Press, 1997; Phillips, K. A., R. L. O'Sullivan, and H. G. Pope. "Muscle dysmorphia." *Journal of Clinical Psychiatry* 58 (1997): 361; Pope, H. G., A. Gruber, P. Choi, R. Olivardia, and K. Phillips. "Muscle dysmorphia: An underrecognized form of body dysmorphic disorder," *Psychosomatics* 38 (1997): 548–557.

Angelique C. Harris

Neck

Neck Lift

A neck lift, or platysmaplasty, is a procedure designed to reduce the loose look of sagging skin in the neck area and under the jawline. Frequently carried out in combination with a facelift, or rhytidectomy, platysmaplasty is performed more frequently on women than men and on adults over, rather than under, the age of forty.

Physician Charles Conrad Miller (1880–1950) of Chicago is credited with developing early platysmaplasty techniques. In the 1920 edition of his book *The Correction of Featural Imperfections*, he described treating a "double chin" by removing excess skin and fat via a long, horizontal incision just under the chin. From the 1940s onward, greater attention was paid to improving methods for tightening the platysma, the broad, fan-shaped muscle running along each side of the neck, extending from the upper part of the shoulder to the corner of the mouth. For example, in 1974, the Swedish surgeon Tord Skoog (1915–1977) argued that cutting away excess skin of the face and neck would not produce sufficient results; he pioneered a technique in which the skin and platysma are detached from the underlying bone and shifted backward and upward as a single unit. In the late 1970s, some plastic surgeons began to use lipoplasty (liposuction) as a method of sculpting the area beneath the chin and jawline. Around the same time, surgeons began to focus their attention on improving the appearance of platysmal "banding."

The history of both rhytidectomy and platysmaplasty are closely tied to the historical development of the profession of plastic surgery. Having gained considerable notoriety for the reconstructive "miracles" performed on men whose faces were injured during World Wars I and II, plastic surgeons began to carry out increasing numbers of facial rejuvenation operations in the postwar years. The rising rate of facial cosmetic surgeries was linked to a number of factors, including the availability of antibiotics for treating postsurgical infection, the development of new surgical techniques during the war years, and the growing realization, among plastic surgeons themselves, that the future of their specialty was less than certain. Regarding the final point, although many plastic surgeons operating during World

Wars I and II voiced confidence that the large numbers of civilians injured in accidents and fires would keep their specialty busy throughout the late 1940s, 1950s, and 1960s, they soon realized that patients were too few to occupy the large numbers of plastic surgeons that the wars produced. The specialty responded to this shortfall by initiating a new trend: marketing particular medical techniques and technologies to specific groups. The first issue plastic surgeons targeted was aging, and their first audience, middle-aged, middle-class women.

This strategy was clearly one whose time had come. In the United States, the generation which came of age following the World War II married and had children younger than had their parents or grandparents. They also lived longer than previous generations. These changes produced a large segment of the American population who, by the age of fifty, were healthy, affluent, finished with child rearing, and ready to enjoy the better things in life. Unfortunately for them, this was also a period when the American fascination with youth, as a necessary component of beauty (particularly for women), had grown particularly acute. In this context, new cosmetic surgical techniques offered aging, middle-class women a new means for fulfilling traditional notions of female responsibility for looking one's (youthful) best.

Like facelifts, platysmaplasty has become one of the procedures commonly employed in the pursuit of ageless beauty, features of which include facial symmetry, high cheekbones, and an angular jaw-neckline. As the human face ages, a relatively consistent process of anatomic change occurs. Specifically, there is a loss of tone in the elastic facial fibers and a corresponding sagging of the skin and soft tissues of the face and neck. Aging of the lower face also often includes drooping, or "jowling," of the soft tissues of the chin and cording of the muscles in the front of the neck. Based on a particular patient's features, cosmetic surgeons today employ a variety of techniques aimed at improving the appearance of the neck and chin. Frequently, incisions are made immediately in front of and behind the ear, through which fat is removed from the jawline and the platysma is excised to gain the appearance of greater tightness. The surgeon may also make an incision directly below the chin in order to remove the fat bulge there; as an alternative, he or she may use a small cannula to suction fat from beneath the chin. Another technique involves cutting into the bend between the chin and the neck and suturing the platysma in order to reduce platysmal banding.

Whatever the procedures used, the goal of platysmaplasty remains the same: to restore the appearance of a youthful neck. Various authors have sought to define the specific neck characteristics perceived as "youthful." Most commonly, these are defined as: a distinct mandibular border (jawline) from the front of the chin to the angle between the chin and neck; a clear depression immediately below the hyoid bone (that is, the U-shaped bone at the root of the tongue); a slightly visible thyroid cartilage bulge, or Adam's apple; a visible front border of the ropey muscle that runs from the rounded bump behind the ear to the joint between the collarbones (the sternocleidomastoid muscle); a cervicomental angle (or the angle between the middle of the under-chin and neck) of 105 degrees to 120 degrees; and, a 90-degree angle between the chin and neck.

Further Reading: Ellenbogen, Richard, and Jan V. Karlin. "Visual Criteria for Success in Restoring the Youthful Neck." *Plastic and Reconstructive Surgery* 66, 6 (1980): 827–838; Haiken, Elizabeth. *Venus Envy: A History of Cosmetic Surgery.* Baltimore, MD: The Johns Hopkins University Press, 1997; Rohrich, Rod J., Jose L. Rios, Paul D. Smith, and Karol A. Gutowski. "Neck Rejuvenation Revisited." *Plastic and Reconstructive Surgery* 118, 5 (2006): 1,251–1,263.

Debra Gimlin

Neck Rings

The necks of humans have been adorned with various jewels throughout history, and women in particular have tried in many ways to elongate their necks or to create the illusion of a long neck. Women in history who have been known for their long and beautiful necks include the Egyptian Queen Nefertiti. However, some neck-adornment traditions modify the body permanently in order to create the appearance of a long neck. Traditional neck rings worn by some tribes in Southeast Asia and Africa appear to dramatically elongate the neck and confer status to the wearer. Indigenous neck rings are worn for beautification, but also for protection and maintaining cultural heritage. Once widespread in some areas, the practice has diminished considerably under pressure from colonial and postcolonial governments. Indigenous women in Southeast Asia and Africa continue to fascinate Western tourists and anthropologists because of the restrictive rings they wear.

Brass neck rings worn in the traditional style of the Padaung tribe of northern Thailand. Courtesy of Corbis.

Southeast Asia

The Karen or Padaung tribe lives today in the Mae Hong Son area of northwest Thailand. The term "Padaung," or "long necks," refers to their elongated necks due to the multiple brass rings they wear. In early colonial Burma, long before it was officially called Myanmar, the Karen tribe, mainly consisting of farmers in the Irwaddy and Salween Valleys, was considered a faithful servant of the British. However, after the British left Burma in 1948, many Karen villages were attacked and burned by the Myanmar army. Many Karen escaped to refugee camps in Thailand. There are some families of the Burmese Karen tribe that have now lived in Thailand for almost 300 years. Many have dual citizenship.

Their rituals and traditions are so ancient that many Padaung claim to have forgotten them. Among the many Padaung customs are the practices of adorning the women of the tribe through fashioning brass neck rings to elongate the neck. The practice of stretching the neck as a social marker starts for many girls at a very young age, with the placing of between three and eight brass rings on the neck. A white paper sheet is sometimes worn underneath to add comfort and act as a barrier between the metal and the soft neck muscles. By the time a woman is an adult, she may have between twenty, thirty, or even forty brass rings on her neck, up to eighteen inches high.

The neck is not actually stretched, but the appearance of an elongated neck is achieved by permanently compressing the collarbone and the shoulders. Neck muscles may be stretched somewhat as well. While posture is certainly affected, there are various other effects these rings have on the neck and body of those who wear them. Movement of the head and neck is restricted. The weight of the rings can crush the collarbones and even break ribs. More often, however, the vertebrae in the neck can become permanently altered. The trauma to the neck from this custom can cause hematoma, or internal blood hemorrhage in the neck. The skin under the rings is also at risk for infections if left unwashed. Removal of the rings is rare, but reasons include for medical examination, to change coils, and sometimes as a punishment for adultery. Some women may choose to remove their rings. However, the muscles in the neck can be extremely weak after many years of adapting to the rings, so even a woman who might want to remove them may be discouraged from doing so. The neck can become bruised and discolored, which will keep an unmarried woman from removing her rings.

There are many different explanations of why the Karen-Padaung practice this custom. Some are confirmed by the women of the Karen tribe, while others seem to be pure speculation. Some of their own stories include that, historically, the rings brought protection from tigers, and made the women unattractive to slave traders. Women are expected to wear neck rings, and they are seen as necessary for finding a mate. Some Burmese refugees in northern Thailand continue to stretch out their necks because of economic reasons. The stretched necks bring a significant number of tourists.

There is a considerable history of Western fascination with and exploitation of the Padaung. Some women were even taken to England in the 1930s and paid to be displayed in the Bertram Mills Circus as ethnological freaks. A Belgian comic book told tales of these women, called *Les Femmes Giraffes*. Even today, Western marketing strategies have included the image of neck elongation as signifying exoticism, sexuality, and attraction. A Christian Dior perfume bottle and advertising used the image of a woman and brass rings. The Icelandic singer Björk also used imagery referencing the Karen tribe on her compact disc cover designed by Alexander McQueen. Many Karen no longer practice this custom. But for those who continue to stretch their necks, there are photographers, tourists, beauty contests, and other carnival events that make this practice a profitable choice. While refugees with long necks seem to be exploited without complaint, the choice is lucrative for tribespeople who rely on tourism.

Africa

The Ndebele tribe is one of the smaller groups living in South Africa, mostly in Mpumalanga and around the Northern Province. They are associated with the Nguni people. The Ndebele, along with other African tribes, often wear neck rings made of copper and brass. Traditional dress also includes colorful caps or head coverings, aprons, and other cloths made of bright and geometically patterned fabrics.

Young girls of the Ndebele tribe wear a ring around their neck; they can marry as early as fifteen, and start wearing the neck rings at that time. Ndebele girls and women also sometimes wear rings on their legs, arms, and waists. Some rings are formed with grasses or cloth material and covered with beads; married women wear metal rings of brass and copper, sometimes adorned with beaded hoops. "Isigolwani," a type of grass rings, are worn only by married women, but Ndebele women have also traditionally worn, as a symbol of their wealth, thick brass rings on their necks and legs. If a woman is married, but does yet not have a home built by her husband, it is common for her to wear a similar neck adornment called a "rholwani." After the home is built, this rholwani is then replaced by her husband with isigolwani rings, or "idzil," made of grass or copper.

In ancient Ndebele traditions, some wives wore brass and copper rings around the neck, arms, and legs. They were a symbol of her loyalty to her husband, and could be removed only after her husband died. Those rings were called "idzila," and were a sign of her husband's wealth. A similar tradition was common in the ancient Ndebele, or Ndzundza tribe of Zimbabwe. Traditionally, the rings would be made by her husband for a woman as a wedding present. These rings were called "dzilla."

Neck rings were once part of the adornment of royal Zulu courts, and varied by the status of the wearer. Both women and men wore copper rings. The chiefs wore "izimbedu," and the "iminaka" were worn by higher members of the royal court. The copper was imported from Delagoa Bay, a port controlled at different times by the English, Dutch, and Portuguese, and often in solid blocks called "umdaka." The metal was transformed into the rings, balls, or plates used to beautify the court and the wives of the chiefs. There were many types of these rings, some made with copper balls called "izindondo." They were a symbol of royal beauty, and continue to represent the elite in South African culture.

The Western world has capitalized on the look of the Ndbele as well as the Karen tribe. A Princess of South Africa Barbie doll, created by Mattel, wears the "izixolwana" rings. The doll is dressed in the brilliant colors associated with the tribe, bright red, blue, and green. While there are many tribes in South Africa, with different tribal costumes and traditions, the Ndebele image was chosen to represent this region by Mattel, whose dolls are famous worldwide.

Further Reading: Delang, Claudio. *Living at the Edge of Thai Society: The Karen in the Highlands of Northern Thailand.* New York: Routledge, 2003; Karen Human Rights Group (KHRG), "Dignity in the Shadow of Oppression: The Abuse and Agency of Karen Women under Militarisation," http://www.khrg.org/; Kennedy, Carolee G. "Prestige

Ornaments: The Use of Brass in the Zulu Kingdom." *African Arts*, vol. 24, no. 3, *Special Issue: Memorial to Arnold Rubin, Part II* (July 1991): 50–55, 94–96; Levine, Laurie. *The Drumcafe's Traditional Music of South Africa*. Johannesburg: Jacana Media, 2005; Levinson, David H. *Encyclopedia of World Cultures: South Asia, Vol. 3*. New Haven, CT: Yale University Press, 1992; Lewis, Paul. *Peoples of the Golden Triangle*. London: Thames & Hudson, 1998; Marshall, Harry I. *The Karen of Burma*. Bangkok: White Lotus, 1997 [1922]; McKinnon, John. *Highlanders of Thailand*. Oxford: Oxford University Press, 1983; Mirante, Edith. *Down the Rat Hole: Adventures Underground on Burma's Frontiers*. Hong Kong: Orchid Press, 2005; "Ndebele," http://www.tribes.co.uk/countries/south_africa/indigenous/ndebele.

Margaret Howard

Nipple

Nipple Removal

Quite rare, nipple removal is the removal of the nipple, areola, and sometimes the skin surrounding the nipple for medical or aesthetic reasons. The nipple is a small mound of skin surrounded by the areola. Nipples contain lactiferous ducts designed to funnel milk. Nipples, also known as the mammary papilla, are found in mammals. The nipple consists of a number of nerves and, as a result, can be quite sensitive. The female nipple tends to be larger than the male nipple and is attached to the mammary glands located in the breast. Both males and females are born with nipples, though nipples have no real function in men. Many males are actually able to lactate, or produce milk, but not enough for nursing infants. Some babies are born with lactating nipples; however, this often goes away within a few days of birth. There are three main reasons for the removal of the nipple: aesthetic body modification, the consequences of cancer, and the removal of a supernumerary nipple.

Body Modification

Body modification can take many forms. In addition to piercings and tattoos, there are more radical forms such as dermal punches, scarification, implants, branding, and nipple removal. Most often, for aesthetic body modification, nipples tend to be pierced; more rarely, in extreme body modification, the nipple can be partially or completely removed. Medical doctors do not routinely perform nipple removal for body-modification purposes. A body-modification artist usually performs this procedure.

During nipple removal, most practitioners use a clamp that is placed over the nipple and often the entire areola as well. The nipple and areola are sliced away. The nipple can also be removed by simply using a sharp tool to cut it away without a clamp. Once cut away, bandages are placed over the tissue. Some body-modification artists recommend that their client go to a doctor for antibiotics to prevent infection. It will take several months to a year for the area to heal properly, leaving scar tissue.

Nipple removal can be a particularly dangerous form of body modification. In nonsurgical settings, only topical anesthetics are used, and as a result, it can be quite painful, and the body can go into shock. There is the possibility of large amount of bleeding. Infection can easily set in as well as difficulty in healing.

Nipple removal is controversial. The legality of the nipple removal solely for aesthetic purposes is in question, and may vary based on location as well as the credentials of the practitioner and the tools used in the procedure. For the most part, those who perform it can face prosecution for practicing medicine without a license. Therefore, the practice of this form of modification is often underground, performed in a secretive manner.

Enthusiasts of this form of body modification tend to be male. This procedure is rarely performed on women, because nipple removal can result in mastitis, or the infection of breast tissue. The ducts in women's breasts that enable breastfeeding can allow for the infection to spread deep within the breast tissue.

Supernumerary Nipple

A supernumerary nipple occurs when there is at least one additional nipple located on the body in addition to the two located in the expected positions on the chest. Such nipples are also known as a "third nipple" or a "superfluous" nipple. Usually, people with these nipples were born with them. Sometimes, they can grow during a person's lifetime, but these instances are quite rare.

Supernumerary nipples are small abnormalities and are often believed to be small moles or birthmarks on the body, as they are normally not accompanied by an areola, are quite often small, and not completely formed. However, some people have been known to have patches of hair surrounding their supernumerary nipple. In some very rare cases, these nipples are accompanied by breast tissue and mammary glands that can produce milk; these are called polymastia. Supernumerary nipples tend to be located along the embryonic milk line, or the area where breast tissue normally forms in mammals. However, supernumerary nipples can be found elsewhere on the body, such as the forehead, stomach, thigh, or foot. These nipples are usually found below the normal nipples on the chest. Reports of supernumerary nipples have dated back to ancient Rome and Greece; anthropologists have found images depicting what is believed to be a third nipple. For example, Artemis, daughter of Zeus and the twin sister of Apollo, is depicted with multiple breasts and nipples in the ancient statue of the Lady of Ephesus. (Some recent scholars now identify them as bulls' testes rather than breasts.) Her temple in Ephesus, Turkey, has been named one of the Seven Wonders of the Ancient World. But supernumerary nipples occur in less than 2 percent of the population. Most often, people do not even realize that they even have a third nipple. Studies conflict over whether men or women have higher rates, but most tend to agree that men are slightly more likely to have supernumerary nipples than women. Supernumerary nipples occur in all racial and ethnic groups. Most people do not get their supernumerary nipples removed, and when they are removed, the reason is usually simply cosmetic. Often, the nipples that are removed are the ones that are very large and are accompanied by breast tissue. A surgeon or a general doctor removes these nipples.

Supernumerary nipples most often go unnoticed, but when revealed, the reactions to these nipples can vary. Some people take great pride in their third nipple, proudly piercing or tattooing around it to bring attention to it. There have also been instances of people in some cultures reacting very negatively to supernumerary nipples, with some believing them to be a sign of witchcraft or some other sort of sorcery or the work of the devil, calling these the "devil's mark." For example, upon her downfall and execution, Anne Boleyn, wife of King Henry the VIII, was deemed a witch in part because she was accused of having a superfluous nipple.

Cancer

Sometimes nipples are removed as a result of cancer. Breast cancer sometimes necessitates a mastectomy, a surgical operation in which the entire breast is removed if the cancer has spread throughout the breast tissue. There are other instances of cancer of the nipple and the milk ducts, which is called mammary Paget's disease. This is an extremely rare form of breast cancer that only affects 1–2 percent of all those with breast cancer. One way to cure this rare form of cancer is to remove the nipple. Often the most challenging part of reconstructive breast surgery is the creation of realistic-looking nipples.

Further Reading: National Cancer Research. www.cancer.gov/cancertopics/factsheet/sites-types/gadgets-breast.

Angelique C. Harris

Nose

Cultural History of the Nose

The human nose, a cartilaginous protrusion from the face, bares the nostrils and plays central roles in both olfaction (smelling) and respiration (breathing). While important medically and functionally, the nose has been given a great number of cultural and symbolic meanings throughout time. Due at least in part to its central location on the face, the nose has been the focus of a great deal of aesthetic attention, both in art and in life. Many arguments and conjectures as to the dimensions of the "perfect" nose have been posited and contested. A variety of cosmetic measures—most notably surgical intervention—have appeared throughout history as a means by which one could achieve a more "perfect" nose.

The nose has also been assigned a major role in the historical project of classifying human beings into groups of distinct races. Several attempts have been made to use nasal proportions or measurements as evidence for a variety of human differences. In fact, many anthropologists have used nasal indices as tools for grouping populations together well into the twentieth century. While some attempts at making classifications based on nasal proportions have fallen by the wayside (or were never popularly accepted), certain typifications have maintained some degree of popular currency, most notably with regard to Negroid and Jewish noses. In fact, at varying points in history, the boom in rhinoplasty (surgical alteration of the nose's appearance) was fueled largely by Jews wishing to avoid the stigma afforded them by virtue of the shape of their noses. Rhinoplasty has also been used by people wishing to avoid the stigma of a nose damaged by syphilis.

Throughout human history, many rituals and customs have centered on the nose. A number of cultural groups have been known to pierce the nose for a variety of reasons. The placement of the piercing as well as the object placed in the opening are indicative of the piercing's cultural relevance. For example, in some cultures (such as in parts of India), a piercing in the left nostril is associated with the health of the female reproductive system. In other cultures, a bone placed through the septum was said to ward off evil spirits and ill health. Nose piercings have also become common among

Eleanor Roosevelt rubbing noses with a Maori woman from New Zealand. Courtesy of Library of Congress, LC-USZ62-64437.

many subcultures in the modern West, largely for aesthetic reasons. Nose piercings, especially nostril piercings, have become increasingly common, and are now even arguably part of the mainstream. Rubbing noses together or smelling a loved one at close range have also been used as customary displays of affection or high regard in a variety of cultures.

Defining the Nose

The human nose can be defined as the protruding part of the face that bears the nostrils. Its primary functions include its role in respiration (providing an entryway/exit for air, along with the mouth) and olfaction (the nose and nasal cavity house a number of olfactory sensory neurons, the starting point for our sense of smell). The nose is given its shape largely by the ethmoid bone and the cartilaginous nasal septum. While the nose itself is a universal feature among human beings, it can take a variety of sizes and shapes and maintain normal function. The various sizes and shapes of the nose, however, have taken on a variety of cultural functions and meanings.

The Ideal Nose

For much of Western history, there has been a fascination with the concept of an "ideal nose." Ancient Greek and Roman artists and sculptors fashioned their artistic subjects with noses that adhered to very particular aesthetics, even mathematical standards of placement, width, length, and shape. One particular historical measure of a nose's aesthetic value is the nasal index. The nasal index, a measure initially described in the 1700s and commonly used by anthropologists well into the twentieth century, is a numerical representation of a nose's relative broadness. The nasal index number represents a ratio of a nose's width to its height. The lower the index number, the narrower the nose. Generally, narrower European noses were taken to be the aesthetic standard or "default." The nasal index was so commonly accepted among anthropologists that many used it as a major criterion for creating racial groupings. Another related measure of a nose's aesthetic value is the facial angle. The facial angle, also defined during the 1700s, more or less represents the angle that would be created by drawing a line from the forehead through the tip of the nose, and connecting this line to a line drawn horizontally outward from the jaw or spine. Anthropologists have observed that the average among Europeans is approximately 80 degrees. Again, this was assumed to be the norm or ideal. Smaller, more acute angles (such as would be made by flatter noses) were considered less appealing. Among ancient Greek and Roman artists, the aesthetic ideal appeared to be close to 100 degrees. Of course, for facial perfection to be achieved, the facial angle had to be considered along with the nasal index. Indeed, a nose with an "ideal" facial angle could be considered large and unseemly if paired with too high of a nasal index.

Interestingly, abstract aesthetic ideals such as those expressed in art often vary greatly from human reality. Roman sculptors systematically created statues with facial angles of 96 degrees, and Greeks often used facial angles of 100 degrees. This is despite the fact that most noses in their populations would have had facial angles of roughly 80 degrees or less. Today, contemporary studies of Caucasian noses suggest that few if any people possess facial angles or proportions that match what most plastic surgeons would consider the ideal. Furthermore, even fewer Caucasians actually seek to achieve that abstract ideal when opting for rhinoplastic procedures.

Typing and Racialization

When the idea of genetically distinct races was en vogue, many anthropologists relied on measurements of noses as criteria for membership in a particular racial group. Although anthropologists conceded that there was great individual variance in nasal indices even within relatively homogenous populations, they argued that distinct racial groups would fall around some mean nasal index number. By and large, anthropologists used the following nasal index groups to roughly divide the human population into broad racial categories:

Nasal Index #	Category	Description
Minimum–70	Leptorrhine	Narrow nose (i.e., Caucasoid)
70–85	Mesorrhine	Medium nose
85–100	Platyrrhine	Broad nose
100+	Hyperplatyrrhine	(i.e., Negroid)

Anthropologists initially assumed that nasal indices (or nose shapes in general) were biologically heritable characteristics that were intrinsically linked to other racialized characteristics. However, throughout the twentieth century, scientists uncovered evidence suggesting that nasal indices were linked to climate rather than to some "race gene." Evidence suggests that higher nasal indices are better suited to hotter, more humid climes, whereas lower nasal indices are suited for colder, dryer areas. It has been theorized that this is related to the fact that air must be sufficiently warm and moist in order for the lungs to absorb it properly. In warmer, more humid climes, the body has little or no work to do in order to prepare the air for gas exchange in the lungs. In colder areas, however, the body needs more surface area (that is, longer nostrils or nasal passages) to sufficiently warm the air. Several anthropologists throughout the twentieth century have suggested that evidence for this theory is strong, given that populations' average nasal indices tend to gradually change over time when the population migrates to an area with a vastly different climate.

Another nasal classification system that for a time appeared to be inextricably linked to race was nasology. Nasology was a nineteenth-century spin-off from phrenology, a pseudoscience that linked a variety of personality traits and character tendencies to the shape of the human head. Nasology claimed that one could deduce certain character traits (including criminality) from the shape of a person's nose. Initially intended as a joke at phrenology's expense, nasology came to have some degree of popular appeal. This is particularly true in terms of the Nubian/wide nose and its relation to criminal behavior or other negative stereotypes associated with Africans and their descendants.

Similar to this concept is the idea that Jews are distinguishable by their noses, which are said to be disproportionately large and hooked. This idea was particularly strong when Jews were viewed predominantly as a "race" of people, as opposed to a religious group comprised of several divergent ethnic groups. At various points in history, it was commonly believed that the "Jewish nose" was not only a telltale marker of Jewish ancestry, but a cue to the fact that a person with such a nose was likely to be shifty, stingy, or any number of negative attributes once commonly associated with the Jewish population.

Rhinoplasty

Rhinoplasty is the surgical alteration of the nose and its surrounding area. Generally considered plastic or cosmetic surgery, nose jobs can be performed to change or improve the overall function of the nose (in the case of severe deformities or disfigurement resulting in difficulty breathing) or to alter its aesthetic appeal. Evidence suggests that some form of rhinoplasty existed as early as 500 BCE, but purely cosmetic rhinoplasty as we know it today was first practiced in the 1800s.

Due to the nose's central location on the face and ready visibility, many people have a strong desire to correct facial deformities or birth defects that affect the nose for functional as well as aesthetic reasons. However, a significant proportion of cosmetic rhinoplasties are performed in order to mask or alter the appearance of racialized nose shapes. At varying points in history, Jews have been some of the largest consumers of rhinoplastic procedures, many openly citing not wanting to be stigmatized for their "Jewish noses" (or needing to "pass" as gentiles for safety reasons). African Americans also have gone under the knife to avoid the stigmatization of stereotypically Negroid nasal dynamics—particularly broad noses or large nostrils. Similarly, nose jobs are extremely popular in some Asian countries, as many people seek to achieve a "more Western" appearance.

In addition to racial and ethnic assumptions related to the appearance of the nose, the state of the nose has been linked to morality and health in a variety of nonracialized ways. This is especially true with regards to syphilis and cocaine use. Because the visible part of the nose is largely given its shape by cartilage and other delicate tissues, this part of the body is highly susceptible to damage and erosion from disease or injury. Syphilis, especially in the later stages of its virulent form or from birth in congenital cases, has been known to cause damage to this delicate nasal tissue (and even the surrounding bone). In eras during which syphilis was widespread—and before effective treatments were readily available—the syphilitic nose was a marker of immorality and evocative of shame. As such, people would go to great lengths to reconstruct their syphilis-damaged noses.

Johann Friedrich Dieffenbach (1792–1847) was one of the central figures in nineteenth-century facial surgery. The chief goal of his practice was to correct and hide the missing or eroded nose, allowing those affected with such a disfigurement to pass as healthy and whole. Many of his patients were suffering from syphilis, either in its virulent or congenital forms. The advances made in nasal reconstruction by Dieffenbach and his students were so great that many modern rhinoplastic procedures are still rooted in his efforts. Dieffenbach once suggested using a gold implant to reconstruct the bridge of a missing nose. Although ultimately unsuccessful in this endeavor, his experiments heralded a new wave of experimentation in facial reconstruction. Along with his colleagues and students, Dieffenbach also revitalized the use of flaps of skin from various body parts to reconstruct the fleshy parts of the nose, a procedure purportedly first used in India as early as 600 BCE. Paramedian forehead flap rhinoplasty is still in widespread use today.

The social stigma associated with a syphilis-damaged nose at earlier points in history bears several parallels with the stigma associated with the cocaine-damaged nose in more recent years. Cocaine is a drug processed from the coca plant. Cocaine (but not as crack or freebase) is generally found in powder form and is commonly ingested via inhalation through the nostrils. Snorting cocaine allows the drug to be readily absorbed by the numerous blood vessels in the nasal cavity. Due to the chemical compounds released at the point of contact when cocaine breaks down, prolonged cocaine use can completely degrade the cartilage of the nasal septum, causing the nose to appear markedly emaciated or deformed. As with syphilis earlier in history, the appearance of a cocaine-eroded nose is an easily read cultural signal of perceived immorality and shame. As such, many people readily seek out rhinoplastic procedures to reconstruct cocaine-damaged noses, both to restore decreased functionality and to avoid social stigma.

Rituals and Customs

The nose has been the site or central focus of numerous cultural rituals throughout human history. Among the most familiar of these rituals or customs is that of piercing. Many societies ascribe some sort of cultural significance to the piercing of one's nose. To create a piercing, a hole is made in one of the surfaces of the nose—generally using a sharp object such as a bone or needle—and a piece of jewelry is inserted. This jewelry can range from hoops or studs made of precious metals, pieces of bone, stone, wood, or precious jewels. While many piercings are done for solely aesthetic reasons, some have some broader cultural significance. For example, many ethnic groups throughout the Indian subcontinent pierce the left nostril of all females because of the left nostril's association with female reproductive health in Ayurvedic medicine. Still other cultures in the Indian subcontinent and the Middle East regard a small hoop or jeweled stud worn in one's nostril as a marker of feminine beauty, marriageability, or social standing.

Septum piercings also hold great cultural and religious significance for several cultures. Most prevalent among tribal societies in the Indian subcontinent, certain parts of Africa, South America, and Polynesia, septum piercings are not only seen as aesthetically pleasing, but are used to ward off evil spirits and ill health by symbolically sealing off the respiratory pathway.

A variety of nose piercings are also common in the modern West. Such piercings are largely done solely for aesthetic purposes. However, for some, facial piercings of any sort represent the bucking of authority and dissent from "traditional" rules for social acceptability. As such, for some, it has come to be viewed as a modern rite of passage.

Aside from decoration or bejewelment, the nose can be the focus of ritual via its role in showing affection. The phenomenon known in the West as "Eskimo kisses" (itself something of a misnomer) is the term used to describe affectionately rubbing one's own nose against the nose of a loved one. This term or idea stems from the fact that various tribal groups have historically not practiced kissing—at least not in the modern Western sense according to the observations of anthropologists. Instead, in such cultures, people may press or bump their noses together as a display of affection. Alternately, people may sniff—sometimes vigorously—the person that they are greeting at very close range, which to some may resemble kissing. Such practices were historically most common among tribal populations in parts of Asia, Polynesia, the American Northwest, the Indian subcontinent, and parts of Africa. Ironically, a number of cultures that have practiced such greeting or expressive rituals and customs have also historically invoked the cutting off of one's nose as a punishment for a variety of crimes.

Notable Noses

Given the central importance of the nose to so many cultures, it is not surprising that many individuals have gone down in the annals of history for the notoriety of their noses. Sixteenth-century Danish astronomer Tycho Brahe (1546–1601) was notable for, among other things, having lost part of his nose in a duel. He purportedly wore a prosthesis made of an amalgam of silver and gold for the remainder of his life. The nose of the character Pinocchio, the wooden puppet first appearing in 1883, got longer and longer the more lies he told. Jimmy Durante (1893–1980), twentieth-century musician, comedian, and entertainer, was similarly well known for his nose. Its large size was frequently the butt of his own comic routines, earning him the nickname "Schnozzola," among others. His likeness appeared in cartoons and comic routines throughout the twentieth century. Fellow twentieth-century entertainer W. C. Fields (1880–1946) also has been parodied in numerous cartoons for the size and shape of his nose. Barry Manilow (1943–) and Barbra Streisand (1942–) both have been the subject of much commentary regarding the sizes of their respective noses and their unwillingness to surgically alter their appearances. Pop-music star Michael Jackson (1958–) is perhaps as notable for the state of his nose as he is for his music career. Jackson and his sister LaToya have both undergone numerous rhinoplastic procedures, rendering each of their noses unrecognizable and significantly altering the appearance of their respective faces. Fictional characters Cyrano de Bergerac, C. D. Bales (a character played by Steve Martin [1945–] in the movie *Roxanne*, a modern adaptation of the story of Cyrano de Bergerac), and Gonzo (of Muppets fame) were all notable for the exaggerated disproportionality of their noses. Fictional character Saleem Sinai, protagonist in Salman Rushdie's novel *Midnight's Children*, possessed an extremely large nose that was somehow linked to a variety of telepathic abilities. *See also* Brain: Cultural History of the Brain; Face: Michael Jackson; Nose: Nose Piercing; and Nose: Rhinoplasty.

Further Reading: Davies, A. "Re-survey of the Morphology of the Nose in Relation to Climate." *Journal of the Royal Anthropological Institute of Great Britain and Ireland* 62 (1932): 337–359; Eichberg, Sarah L. "Bodies of Work: Cosmetic Surgery and the Gendered Whitening of America." PhD diss., University of Pennsylvania, 1999; Gilman, Sander. "By a Nose: On the Construction of 'Foreign Bodies.' "*Social Epistemology* 13 (1999): 49–58; Gilman, Sander. *Making the Body Beautiful: A Cultural History of Aesthetic Surgery*. Princeton, NJ: Princeton University Press, 2000; Hopkins, E. Washburn. "The Sniff-Kiss in Ancient India." *Journal of the American Oriental Society* 28 (1907): 120–134; Jabet, George. *Notes on Noses*. London: Harrison and Sons Publishing, 1852; Leong, Samuel, and Paul White. "A Comparison of Aesthetic Proportions Between the Healthy Caucasian Nose and the Aesthetic Ideal." *Journal of Plastic, Reconstructive and Aesthetic Surgery* 59 (2006): 248–252; Stirn, Aglaja. "Body Piercing: Medical Consequences and Psychological Motivations." *The Lancet* Volume 3b1, Issue 9364 (2003): 1205–1215.

Alena J. Singleton

Nose Piercing

Nostril and nasal septum piercing date back thousands of years and span several cultures. Much of the evidence of nose piercing throughout history has been derived from figurines, artwork, archaeological sites, and, more recently, anthropological accounts.

One of the most ancient instances of nose piercing comes from the Samarran tradition that existed from 5300 to 4000 BCE, in what is now central eastern Iraq. Terra-cotta figurines of women wearing jewelry, including nose studs, have been found, implying that women pierced their noses. Nose studs have also been found among the Samarrans' southern neighbors, the Ubaid (5000 BCE).

As of the early 1900s, nomadic Bedouin women were still piercing their nostrils. A study of Egyptian bedouin from the 1930s shows a photograph of an unveiled woman who is wearing a large, decorated ring in her right nostril. Apparently, the ring is gold; nose rings of cheap gold were believed to cause women to inflict evil upon those they looked at, especially men and children.

There is much evidence for nose piercing among the ancient people indigenous to present-day Alaska and the Aleutian Islands. Dating back to approximately 4000 BCE, those of the Aleutian tradition wore ivory or bone pins through the nasal septum. Amber and lignite beads were also used to decorate the nasal septum. The Tarya Neolithic peoples (2000–500 BCE) used polished stone and bone as ornaments in nasal septum piercings. Anthropologists have speculated that some piercings were used in rites of passage for boys and girls. In the Kodiak tradition (2000 BCE–1250 CE), crescentic rings were used for nose ornamentation.

Depictions of Eskimos and Aleuts in the late eighteenth century show both men and women with nose piercings. The nose ornaments worn included items placed horizontally through the septum. There are also depictions of beads strung through the septum, sometimes hanging in front of the lips. It was observed in the early nineteenth century, that, during special occasions, strings of beads were attached to simple ornaments. These beads were made from mussel shells, coral, glass, or amber.

Practices varied by region. Some would pierce the septum of a child within the first twenty days after birth following a purification bath. Others would pierce the septum at a later time during childhood, while some waited until after marriage. Usually, the piercing was a special occasion calling for a feast. These practices were already rare by the mid-nineteenth century. It is common to see such customs fade out after contact with the West and colonization by European cultures.

Some groups indigenous to Mexico wore ornaments in nose piercings. There is an Olmec jade pendant dating to the first millennia BCE that depicts a human figure with a pierced nasal septum. In the central Mexican, classic tradition dating back to 200 BCE, the elite classes wore nose pendants. Ornaments often carried specific ranks or offices rather than representing generic status. From roughly 850 to 1500 CE, nose ornaments were made from greenstone, jade, quartz, and onyx. Fifteenth-century accounts from Spaniards record Aztec emperor Tizoc with a septum piercing filled with jewelry made of green stone.

People of the Late Andean Formative tradition (1050 BCE), living in an area along a coastal desert strip of the Pacific Coast in present-day Peru, used gold nose ornaments, although the details in regard to dates are sketchy. Among the Nasca of the southern coast of Peru from 250 BCE to 650 CE, those of an elite status wore nose ornaments. As of the 1990s, the Ucayali of Peru's eastern rainforest were still wearing strings of beads and larger silver pendants through the nasal septum. The members of the

Maruba tribe of the Brazilian Amazon also have their septa pierced and wear strands of beads through the piercing.

Nose piercing has been the most popular in India. The importance of jewelry in India has roots in religion, tradition, and aesthetics that stretch back several thousand years. However, by many accounts, nose piercing dates back only to the fifteenth century CE, when the practice filtered into India from the Middle East, and during the sixteenth century CE under the Mughal dynasty.

There are several kinds of nose rings in India. Various types can be worn in either the right or left nostril or through the septum. Further, the nostrils may be pierced toward the side or middle of the nose; sometimes, all three are pierced, with jewelry worn in each hole simultaneously. Some of these variations are associated with specific regions in India. The ornaments vary from rings to studs; they may be jeweled or plain, with pendants or without, and with chains or without. Size varies along with materials, which include gold or silver with or without additional ornamentation.

A collection of photographs in Karl Gröning's *Decorated Skin: A World Survey of Body Art* (2002) displays women in the 1990s from different parts of India with varying nose ornamentation. A girl from Rajasthan, in the north, wears metal nose jewelry through her left nostril with a string of beads connecting the piercing to her ear. A Muria woman from central India is pictured with facial tattoos and wearing metal ring jewelry in her left nostril. A woman from Kutch (on the northwestern coast) wears left and right nostril piercings and a septum piercing with a beaded pendant.

Nose-piercing revivalists in the United States do not necessarily need to look as far as India for nose-piercing history. Nose piercing also has a history among the indigenous peoples of North America. The Late Hohokam (1050 to 1450 CE) tradition in the present-day southwestern United States made nose plugs of soapstone and argillite that were worn through the nasal septum. Some southern California tribes pierced the nasal septum and wore deer bone through it. According to an account from the late 1700s, the Chumash of the southern California coast near present-day Ventura pierced their septa. During their expedition 1804 to 1806, Meriwether Lewis (1774–1809) and William Clark (1770–1838) brought earrings and nose trinkets to trade with the indigenous people they encountered.

Another famous traveler also dealt with the trade of nose rings. It was recorded in Christopher Columbus' log in 1492 that people indigenous to the Caribbean Islands wore nose rings made of gold and silver. Columbus was able to trade mere trinkets for these precious-metal nose rings.

Nose piercing has been a point of contention in some colonized societies, like the Kuna (also known as the Cuna) of Panama. It was traditional for both men and women to wear ornamentation through nasal septa piercings. The men wore round nose plates of gold that were hung from a septum piercing that covered the upper lip. The women wore circular gold nose rings (*asuolos*) so large that the ring framed the mouth. During the long process of colonization, the Kuna were coerced into adopting Western modes of dress. Long before this, however, it had become taboo for the Kuna to make jewelry. This was due to the invading Spanish conquistadors, who killed for any object of gold. By the early 1900s, it was difficult to find instances of men with nose rings, and women's nose rings became much smaller, not even touching the top of the upper lip. However, there was a Kuna revolt in 1925. Subsequently, the *asuolo* and the *mola* (traditional dress) have become politicized and represent Kuna pride, ethnic identity, defiance of assimilation, and symbols of the Kuna's right to self-governance.

Nose piercing is also a rite of passage in many cultures. Various Australian Aboriginal peoples pierce the septum of young males as a rite. The Pitjanjara pierce the noses of boys to mark the entrance into manhood during dreamtime ceremonies. The Yiwara have the nasal septum pierced before circumcision. Sometimes, the nasal septum is pierced for aesthetic reasons. Some Aborigines insert bone through the septum in order to broaden the nose, which is considered aesthetically pleasing. Aboriginal peoples from Australia to Papua New Guinea use several materials for nose ornamentation: bone, wood, metal, and feathers.

Nose piercing is also practiced among many cultures of Africa. Young girls of the Nuba tribe have their nasal septa pierced and wear nose rings or nose plugs in the piercing. The Hadendawa women of Sudan and the Beja women of North Africa wear nose rings. Muslim Afar women of Djibouti and the Bilen women of Eritrea also practice nose piercing.

Contemporary and Western Nose Piercing

In the West, both nostril and septum piercing became more common around the end of the twentieth century. This can be credited to three main subcultures in the 1970s and 1980s: gay bondage/domination/sadomasochism (BDSM) practitioners, punks, and modern primitives.

Perhaps the group that has been given the least credit for popularizing nasal piercing is the gay BDSM culture. However, some members of this group who became active in the early body-piercing culture in the United States and England had their nasal septa pierced. Two notable examples are Sailor Sid Diller and Alan Oversby (also known as Mr. Sebastian). Punk culture also contributed to nose piercing. Punk-rock youth in the late 1970s used safety pins to pierce their noses and ears.

By far, the group with the greatest influence in contemporary Western nasal piercing has been the modern primitives. Throughout the 1980s, the modern primitives, with Fakir Musafar as the main advocate and the so-called father of the movement, promoted body piercing of all types. After ear piercing, nose piercing became the most common piercing practice among youth in the West. Nostril piercing has become a mainstream practice done primarily for aesthetics, whereas nasal septum piercing and other forms of nose piercing remain countercultural.

The most popular and mainstream form of nose piercing involves piercing one nostril. This is usually done with a small-gauge needle, after which either a stud or ring is inserted. Using a dermal punch or stretching the nostril piercings, often one in each nostril, is becoming more common, though far from mainstream. A nostril piercing can be stretched to accommodate large jewelry or plugs.

Perhaps the second most common type of nose piercing is the piercing of the nasal septum. In most cases, the cartilage is not pierced; rather, the skin between the cartilage and the bottom of the nose is pierced. This piercing may also be stretched to accommodate larger jewelry or a plug. Methods to enlarge the piercing include using a dermal punch, using a scalpel, and slowly stretching the piercing over time. One variation on septum piercing is septril piercing. This is usually done after the septum has been stretched. The front of the nose is then pierced so that the piercing passes from the front of the nose to the stretched septum.

There is also the nasal-tip piercing. This passes from inside the front of the nostril to the center tip of the nose. Sometimes this piercing is called a rhino piercing. The Austin bar piercing is another nose-tip piercing. It passes horizontally through the tip of the nose without entering the septum or the inside of the nostrils.

The nasallang builds off of the look and direction of the genital piercing known as the ampallang. This piercing combines nostril and septum piercings. A single piece of jewelry, most often a barbell, passes horizontally through each nostril and the septum.

Further Reading: *Body Modification E-zine Encyclopedia*, http://wiki.bmezine.com/index.php (accessed May 2007); Kamalapurkar, Shwetal. "Fancy a Dangler Round the Nose?" *Deccan Herald*. January 2006, http://www.deccanherald.com/Archives/jan62006/she115855200615.asp (accessed May 2007); Keeler, Clyde E. *Cuna Indian Art*. New York: Exposition Press, 1969; Peregrine, Peter N., and Melvin Ember, eds. *Encyclopedia of Prehistory*. Vols. 2, 5, 6, 7, and 8. New York: Plenum, 2001, 2002, 2003; Rubin, Arnold, ed. *Marks of Civilization*. Los Angeles: Museum of Cultural History, UCLA, 1988; Tice, Karin E. *Kuna Crafts, Gender, and the Global Economy*. Austin: University of Texas Press, 1995; Ward, Jim. "A Visit to London and Remembering Mr. Sebastian." *Body Modification E-zine*. July 15, 2005, http://www.bmezine.com/news/jimward/20050715.html (accessed May 2007); Vale, V., and Andrea Juno, eds. *Modern Primitives*. San Francisco: Re/Search, 1989.

Jaime Wright

Rhinoplasty

The word "rhinoplasty" comes from the Greek *rhinos*, for nose, and *plastikos*, meaning "to shape." The nose is made up mostly of cartilage, which is easier to work with surgically than bone. Rhinoplasties generally consist of the nose's bone, skin, or cartilage being surgically reduced, augmented, or manipulated to create a different shape. This surgical procedure is commonly known as a "nose job" and is used in cosmetic surgery to change the appearance of the nose. It may be undertaken for aesthetic, cultural, psychological, or social reasons. In plastic or reconstructive surgery it is used to improve the function of noses (often by removing obstructions) that have been damaged through injury or disease. Despite modern-day distinctions between restorative and cosmetic surgeries, the history of rhinoplasty shows how boundaries between reconstructive and cosmetic surgeries have been intertwined and blurred.

Some surgeons use mathematical formulas to determine the "ideal" shape of the nose in relation to the rest of the face, while others rely on artistry

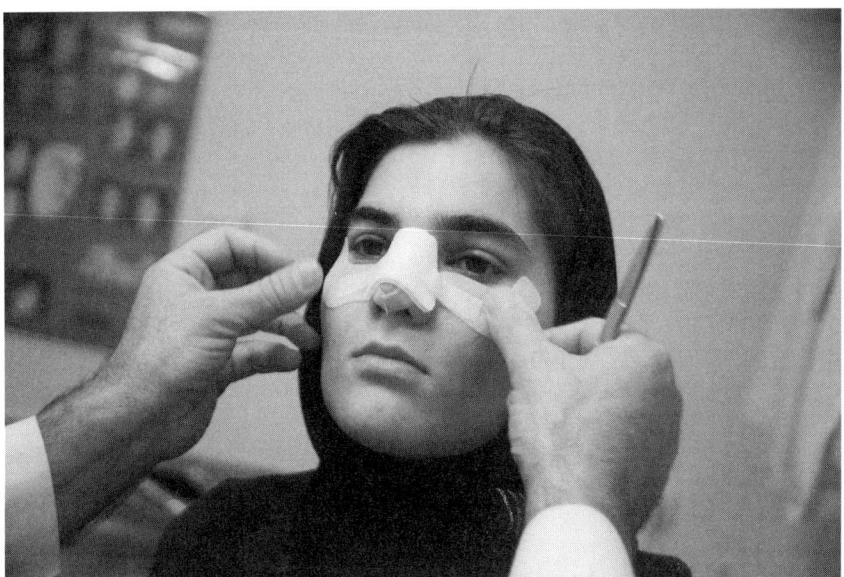

Patient with a bandage on her nose after plastic surgery in Tehran, Iran, in 2005. Nose surgery is popular in Iran for the young wealthier generation. Under Islamic law, women show their faces but keep their hair, arms, and legs covered in public. (AP Photo/san Gary Knight/VII)

or intuition. Noses can be vulnerable to fashion trends, and the shape and size of a beautiful nose has varied considerably over time and in different places. Agnolo Firenzuola (1493–1543 CE), a Renaissance theorist of beauty, said that turned-up noses were unattractive. Now they are the beauty norm.

Rhinoplasty was one of the most common cosmetic surgery procedures undertaken in the United States in 2006: it was the third most popular cosmetic surgery operation for women (after breast augmentation and liposuction), and the most popular operation for men. It continues to be the most commonly sought cosmetic surgery procedure among thirteen- to nineteen-year-olds. Its popularity is rising, with U.S. surgeons reporting increases in instances of the operation of between 9 and 20 percent from 2005 to 2006. Most rhinoplasty procedures are sought by people in the twenty-two to forty-year-old age group.

Early Rhinoplasty

Rhinoplasty is thought to have been developed by Sushruta (ca. fifth century BCE), a physician in ancient India regarded by many as the "father of surgery." He wrote about the operation in his surgical volume *Susrutha Samhita*. Sushruta and his pupils reconstructed noses that had been amputated as punishment for crimes, lost via injuries, or damaged from disease. Sushruta's technique of "flap rhinoplasty," wherein a missing nose was be rebuilt using a flap of skin brought down from the forehead or across from the cheek, is still practiced today using very similar techniques.

In Europe, surgeons also aimed to recreate noses that had been sliced off during war and in swordfights or destroyed by syphilis. Theodoric, Bishop of Cervia (1205–1298) and surgeon Lanfranco of Milan (1250–1315) both described nose rebuilding operations. Syphilis was endemic in Europe by the end of the sixteenth century, and the rise of plastic surgery is closely connected to the spread of this disease, which is mainly sexually transmitted but also can be contracted in utero (congenital syphilis). Many syphilis sufferers, without treatment, will suffer a "rotting away" of parts of the body including the bridge of the nose. (The disease is now easily treated with penicillin.)

The earliest illustrated record of rhinoplasty is from 1597, during the Italian Renaissance, in the book *De Curtorum Chirurgia*. It shows the work of Gaspare Tagliacozzi (1545–1599), professor of surgery at the University of Bologna. Tagliacozzi described rebuilding a lost nose using a pedicle flap graft. He would make two parallel cuts in the flesh of the patient's bicep, loosening the skin in between them and lifting it using bandages. After four days, scar tissue began to form beneath the cuts. Tagliacozzi then severed one end of the flap, leaving the other end attached to the arm. The arm was raised up close to the face, supported by a cast or splint, and, after the area had been abraded (freshly wounded), the skin flap was placed where the new nose would "grow." The patient remained thus—arm attached to face—for twenty days, until the skin could finally be severed completely from the arm. Anesthesia was not yet invented, and altogether this painful and precarious operation took many months of limited mobility with no guarantee of success. It is an indication of the extreme social and cultural stigma of not having a nose that people were prepared to endure such procedures.

Significantly, Tagliacozzi's work was condemned by the powerful Roman Catholic Church. Rebuilding noses was considered to be heretical, unnatural, and immoral, because it minimized the evidence of divine retribution thought to be God's mark on the bodies of "sinners" such as syphilis sufferers. In fact, Tagliacozzi was so despised by the Catholic Church that, after his funeral, his corpse was exhumed and moved to unconsecrated ground, and his innovative surgical work then went ignored and was subsequently forgotten for centuries.

German surgeon Johann Friedrich Dieffenbach (1794–1847) claimed to have invented the modern rhinoplasty, which he would have performed without anesthetic. Anesthesia was developed in 1846, and antiseptic in 1867. Before then, all surgeries were conducted in what would now be considered grossly unhygienic and excruciatingly painful circumstances. Undergoing surgery meant risking death through infection, blood loss, or shock. The advances made in antiseptics and anesthetics were hugely significant in the rise of plastic surgery, and particularly to rhinoplasties, from the late 1800s to the present day.

World War I

Plastic surgery had been considered morally suspect since its inception, but World War I created an opportunity for the development of surgical techniques and for a shift in the public perception of plastic surgery. During

World War I, many soldiers sustained horrific facial trauma in trench warfare, as they were often exposed to gunfire and shrapnel from the shoulders up, and there were more nose injuries than had ever been treated before. In addition, medical advances meant that more soldiers with serious injuries survived battle and needed restorative procedures. Talented plastic surgeons like Harold Delf Gillies (1882–1960), a New Zealander, and Hippolyte Morestin (1869–1919), a French surgeon, honed rhinoplasty techniques in order to rebuild soldiers' noses that had been destroyed in battle. The significant number of facial surgical cases and the public's willingness to support plastic surgery contributed to the development of surgical techniques during this period. Due to its usefulness in treating the injuries of war veterans, plastic surgery achieved some measure of respectability. After the war, Gillies and Morestin set up private practices and began offering rhinoplasties and other cosmetic surgeries to patients seeking them for aesthetic reasons.

Modern Rhinoplasty Methods

Intranasal Rhinoplasty

John Orlando Roe (1849–1915), an otolaryngolotist from Rochester, New York, is credited with developing and performing the first intranasal rhinoplasty in 1887. Roe was aiming to "fix" the "pug" or "saddle" nose belonging to some Irish immigrants. Nose shape and size historically have been linked to weakness or strength of character, and to prove good or bad breeding. Roe categorized various types of noses and attributed personal characteristics to them: a Roman nose signified strength; a Greek nose signified refinement; a Jewish nose signified commercial acumen or greediness; and an Irish or pug nose signified weakness. From very early in the history of modern cosmetic surgery, rhinoplasty was intimately connected to racism: by changing one's nose, one could change one's life, one's prospects, and even one's character or heritage. By diminishing signs of their racial or ethnic origins, many Irish immigrants, especially those in New York, hoped to "pass" and to reduce the discrimination they experienced. However, prior to the intranasal rhinoplasty, they were often left with a telltale scar. Intranasal rhinoplasty was revolutionary because it reshaped the nose without leaving external visible scars; most of the work was done subcutaneously, via the nostrils.

The Contemporary Operation

The intranasal method is still the most widely used rhinoplasty technique today. Performed under general, local, or intravenous-drip anesthetic, incisions are made either inside the nostrils (the "closed" technique) or inside the nostrils and through a cut in the columella, the skin that separates the nostrils (the "open" technique). The soft tissues and skin are separated from the cartilage and bone to allow room for reshaping, augmentation, or diminution. Once revealed, the nasal bones can be broken (a procedure called osteotomy) and then reset or chiseled into different shapes. Cartilage may be trimmed or reshaped. Incisions are usually closed with dissolving sutures but occasionally nylon sutures are used and are removed several days later. Contemporary rhinoplasty operations take about two hours. They are performed in outpatient clinics or in hospitals; patients remain in care from between a few hours to several days. After the operation, a splint is taped over the nose so that it maintains its new shape during healing; sometimes the inside of the nose needs to be packed with bandages. Short-term side effects can include headaches, swelling, numbness, and black eyes. Longer-term complications can include breathing problems, sinus infections, blood clots, and burst blood vessels.

Implants

For nasal augmentations, such as building a bridge, a piece of the patient's rib cartilage (costal cartilage) or ear cartilage (auricular cartilage) is most often used. These are called autogenous cartilage grafts. Specially designed implants can also be made of silicone, Gore-Tex, ceramic, or bioglass. Historically, surgeons have experimented with materials such as gold, silver, lead and aluminum, rubber, paraffin, polyethylene, ivory, cork, stone, and animal bone. However, synthetic materials tend to cause complications, such as migration or extrusion.

Nonsurgical Methods

Nonsurgical rhinoplasty is a relatively new procedure, developed in the early 2000s, where injectable "fillers" such as Restylane, Artefill, and Radiesse are used to change the nose's shape. These substances are injected beneath the skin of the nose to lift low areas or to hide lumps. The results are temporary, lasting between five and twelve months, and no anesthetic is needed.

Ethnicity

The use of aesthetic or cosmetic rhinoplasty has been linked historically to the values and prejudices of the culture in which it is performed. Early modern operations in the United States were deployed to correct small, flat noses, which were linked not only with congenital syphilis, but also with Irish and African features. Now, "ethnic cosmetic surgery" is a growing phenomenon, especially in the United States, where surgeons may claim special expertise in certain groups such as African Americans, Hispanics, or Asians. Cosmetic surgery on "ethnic" noses usually aims to homogenize them to some extent: wide noses are narrowed, flat noses are raised, humps are removed, long noses are shortened, and nostrils are made smaller. People commonly seek rhinoplasties in order to look less Jewish, less Asian, less African American, or less Middle Eastern. This applies in the United States but also across the world: recent similar trends have been documented in Israel, Iran, China, and Japan. Often, rhinoplasties are advertised and sold as "refinements" to the nose; when it comes to ethnicity, this likely means a lessening of ethnic characteristics. Some clinics specifically advertise "Westernization rhinoplasties."

Irish Rhinoplasty

In the late 1800s, Irish "pug noses" could be surgically "Americanized" so that Celts could pass as Anglo-Saxons. At this time, the Irish were thought to be degenerates, and they suffered harsh discrimination. For Irish immigrants in North America, minimizing the signs of one's ethnic heritage facilitated easier social and cultural interactions. Ironically, by the mid-twentieth century, the upturned and small pug nose associated with Irishness became most desirable through cosmetic surgery. In the early twenty-first century, people still generally seek to create smaller, more upturned noses.

Jewish Rhinoplasty

In Germany in the 1800s, Jewish men were able to move relatively freely in the commerce and business worlds, but only so long as they didn't look *too* Jewish. Jacques Joseph (1865–1934), a Berlin surgeon who had trained in orthopedics and had in fact changed his name from the distinctly Jewish Jakob Joseph, performed the first known specifically "Jewish" nose job in 1898. The rise of Jewish rhinoplasty paralleled the anti-Semitism growing in force across the Western world at the turn of the century. In the early 1900s, Joseph refined Roe's techniques in intranasal (scarless) surgery and became internationally famous as the "father of the rhinoplasty."

Jewish rhinoplasty has been performed in Europe and the United States from the mid-1800s. Historians agree that Jews were the first group to really embrace cosmetic surgery, having rhinoplasties commonly from the 1920s. Fanny Brice, a very famous Jewish actress, underwent a well-publicized nose job in 1923 in order to secure better acting roles in Hollywood. "Jewish" noses were thought to be unattractive and un-American, so choosing rhinoplasty was a way to decrease discrimination, increase career opportunities, improve marriage prospects, and allow Jewish citizens to assimilate and blend in with the so-called mainstream.

For European Jews during the Holocaust and throughout the first half of the twentieth century, the desire for alteration of their appearances may well have been for the purpose of survival itself. Looking "too Jewish" is a notion that survives even today. In Israel, rhinoplasty is the most popular of all cosmetic surgery procedures. In the United States, it is not uncommon for Jewish teenagers to seek rhinoplasties. For some, rhinoplasties are a rite of passage and a sign of cultural assimilation, wealth, and prosperity. Critics see such rhinoplasties as a sad indication that Jewish people are still stigmatized.

African American Rhinoplasty

The so-called Negroid nose could be as problematic for African Americans after the Civil War as Jewish noses were for Jews. However, changing African American noses would have been conducted in secret, and the operation's history is undocumented, because surgeons could not risk being seen to be helping black people to pass as white. Indeed, changes to the Negroid nose in the late nineteenth and early twentieth centuries are documented only in terms of white people needing surgery to correct features that happened to seem "too black." There is little doubt, however, that some African Americans undertook rhinoplasties to try to "pass" as white in a racist culture. Today, so-called African American rhinoplasty usually aims to narrow wide nostrils. Sections of the alar (the base of the nostrils) may be removed or sections of the alar groove (where the nostrils meet the face) might be made smaller. In 2006, 62 percent of African American cosmetic surgeries in the United States were rhinoplasties.

There are strong debates surrounding cosmetic surgery and blackness. Critics argue that cosmetic surgery is still used by African Americans to become "less black," that it is a reaction to a racist society, and also that it makes life harder for people who choose not to have cosmetic surgery. Others believe that altering one's appearance has no effect on being black and is merely an individual choice.

Asian and Middle Eastern Rhinoplasty

Asian Americans and Southeast Asians sometimes choose rhinoplasties in which the bridge of the nose is built up, making it appear higher, narrower, and more European. This is achieved through infractures, in which nasal bones are broken and moved inward, and through augmentation of the bridge. Again, these operations are increasingly seen as a sign of belonging to a sophisticated and cosmopolitan Western world, where beauty is a commodity that can be purchased. For the Asian community in the United States (whose cosmetic surgery rose by 58 percent between 2004 and 2005), rhinoplasty is the second most common procedure after blepharoplasty (double-eyelid surgery).

As with noses that are "too Jewish," "too Asian," or "too Negroid," noses that are "too Middle Eastern" are shaved down to more Western dimensions, often to display their owners' cosmopolitanism. In Iran, nose bandages and splints after rhinoplasties are sometimes seen as status symbols, representing, for some, liberal religious and political views.

Rhinoplasty and Culture

Noses hold great symbolic meaning. They are thought by classical Freudian psychoanalysts to be powerful phallic symbols. They are often seen as the most vital indication of a person's race or ethnicity. They play crucial roles in some of our best-known cultural narratives. In fiction, Pinocchio, the wooden puppet character created by Carlo Collodi in 1883 and made famous in Walt Disney's 1940 cartoon, has a nose that grows longer and longer the more lies he tells. Erik, the ghost in Gaston Lerouz's gothic novel *The Phantom of the Opera* (also known as a long-running musical by Andrew Lloyd Webber), has no nose and must wear a facial prosthetic made of papier-mâché. Lord Voldemort, the most evil character in J. K. Rowling's hugely popular Harry Potter series, has no nose at all: merely slits like a snake. Noselessness has traditionally been associated with being evil, and sometimes with being subhuman.

In real life, contemporary pop star Michael Jackson is famous for having multiple rhinoplasties that have radically changed his nose. While Jackson's cosmetic surgery is thought to be extreme, rhinoplasties are commonplace in the entertainment industries, especially among performers. Indeed, there is hardly a Hollywood actor working today who is not known or rumored to have had a nose job.

There are interesting questions around reality and individuality that can be considered in relation to rhinoplasty. Early critics of cosmetic surgery considered whether the person with a nose job was wearing a sort of permanent mask, and what that meant for their authenticity. Contemporary observers have suggested that cosmetic surgeons have promoted a homogenous model of beauty and an Anglicized nose, and have considered the implications for people belonging to distinctive ethnic groups. Whereas once rhinoplasty was taboo and secret, it is often now freely spoken about and is even a status symbol in some communities. "Ethnic" rhinoplasties may be interpreted as a way to conform to Anglocentric cultures, but also may be an indication of an increasingly globalized world in which cultures and ideals of beauty are mixing in ways never previously possible.

Representations

Cosmetic surgery is no longer seen as an indulgence only for the rich or vain. Of all cosmetic surgery procedures, rhinoplasty is probably now the most acceptable, and has become relatively commonplace in wealthy parts of the world. Like all cosmetic surgeries, rhinoplasty has never been simply a physical operation. It has always been linked to a healing or mending of the psyche through physical transformation. Early proponents of rhinoplasty wrote about how it improved the morale of their patients, and nowadays it is commonly accepted that physical changes to this most obvious body part can affect self-consciousness and improve body image. The contemporary belief that the inner self should somehow match one's external appearance is closely tied to the popularity of rhinoplasty, which, depending on context, can make people appear stronger-willed, more sophisticated, more Western, more American, more feminine or masculine, or simply more conventionally attractive.

Before/After

The before/after trope is the principal way in which rhinoplasty has been visually represented in both medical and popular media. Two photos are presented, usually side-by-side. One represents "before" and shows a presurgical nose, while the other shows the postsurgically modified nose. Before/after is used to demonstrate changes wrought by all cosmetic surgery, but it is rhinoplasty that appears to change the face most dramatically, especially when the photos have been taken in profile. In written and spoken media, rhinoplasty is often represented as a relatively minor surgery consisting of "refinements" and "subtle changes;" it can be presented in much the same way as a new hair color or makeup style. This is partly a deception (because rhinoplasty is major surgery), but also, importantly, it is an indication of the extent to which many cultures have embraced the possibility of transforming the body at will. *See also* Face: Michael Jackson; and Nose: Cultural History of the Nose.

Further Reading: American Society for Aesthetic Plastic Surgery, http://www.surgery.org/; American Society of Plastic Surgeons, http://www.plasticsurgery.org; Blum, Virginia. *Flesh Wounds: The Culture of Cosmetic Surgery.* Berkeley: University of California Press, 2003; British Association of Aesthetic Plastic Surgeons, http://www.baaps.org.uk/; Gilman, Sander L. *Making the Body Beautiful: A Cultural History of Aesthetic Surgery.* Princeton, NJ: Princeton University Press, 1999; Haiken, Elizabeth. *Venus Envy: A History of Cosmetic Surgery.* Baltimore, MD: The Johns Hopkins University Press, 1997; Taschen, Angelika, ed. *Aesthetic Surgery.* London: Taschen, 2005.

Meredith Jones

Ovaries

Cultural History of the Ovaries

The ovaries are the two female sexual reproductive organs that are located on either side of the uterus and sit above the fallopian tubes. The ovaries are glands known as the female gonads that produce the sex hormones estrogen and progesterone and control the development of sexual organs and the menstrual cycle. Ovaries are about one-and-one-half inches in size and are shaped like teardrops.

Females are born with approximately one million immature eggs, and each ovary contains cells with the eggs in the center. During childhood, approximately one-half of ovarian follicles are absorbed by the body. By the time a girl reaches puberty and her menstrual cycle begins, only about 400,000 ovarian follicles are left to develop into mature eggs. Starting at puberty, either the right or left ovary produces a single mature egg for fertilization. The eggs mature, and one breaks off through the ovarian wall approximately every twenty-eight days in the process known as ovulation. After its release from the ovary, the egg or ovum passes into the fallopian tube and into the uterus. Although only one egg usually fully matures during ovulation, somewhere between ten and twenty follicles begin the process of maturation monthly, and the excess ovarian follicles are reabsorbed before ovulation occurs. The female hormones produced by the ovaries, estrogen and progesterone, are often considered significant in determining sexual difference.

Discovery of the Ovaries

Aristotle (384–322 BCE) made references to oviducts, and it is possible that he knew of the ovaries. Historians of medicine credit Diocles of Carystus, who lived sometime in the fourth century BCE, with discovering the ovaries; however, in some accounts, they were discovered by Herophilus (335–280 BCE), a physician in Alexandria who is also credited with the first human dissection. The influential physician Galen (ca. 129–ca 200 CE) wrote in the second century CE that Herophilus' description was highly accurate. However, it was not until much later that the role of ovaries in human reproduction was well understood.

The notion of how ovaries were viewed in relation to reproduction reflects the prevailing theories and arguments about the origins of life and human development. One theory known as "preformation" is a preevolutionary theory of human development that posits that the individual starts from material that is already preformed. This theory was supported by scientists who used early microscopes, which were developed in the late 1600s by Antoni van Leeuwenhoek (1632–1723). Van Leeuwenhoek's student, Dutch scientist Nicolass von Hartsoeker (1656–1725), observed sperm under a microscope in 1694. He took the theory of preformation literally and was said to have identified a miniature man on the head of the sperm. In the late eighteenth century, this hypothesis was further explained by Swiss naturalist and philosopher Charles Bonnet (1720–1793), who put forth the theory that all females carry within them all future generations that were individually predetermined. Preformation was a popularly held view, partly because it was understood to be compatible with biblical views of reproduction.

During the seventeenth century, scientists known as "ovists" believed that inside each female egg there was a preformed embryo that contained a future human, that embryos were already perfectly formed in miniature inside each woman, and that they needed only to be given sustenance to grow into a fetus. Ovists thought that sperm was necessary only to stimulate the egg to grow in the womb and that women carried eggs of both genders, which was determined before conception occurred. Those who held an alternative view of conception during the 1700s were known as the "spermisists." They believed that the sperm contained the preformulated materials that were needed to create life and that the egg was merely provided the environment for the fetus to grow. They claimed that the primary contribution of the female to the next generation was her womb.

The theory of epigenesis traditionally runs counter to that of preformation in the debate over how life begins. Epigenesis argues that a fetus starts from material that is unformed and that its form emerges gradually. Aristotle was an early epigenesist, and claimed that life starts and individuals are formed when the body fluids of a man and a woman come together. Aristotle stated that this act brought an individual from a potential to an actual being that continued to grow in the woman's uterus. This gradual development of the human was also supported by Catholic philosophers St. Augustine (ca. 354–430) and St. Thomas Aquinas (ca. 1225–1274), who believed that becoming a human did not occur until about forty days after the mix of male and female fluids, at which time a soul developed. In the eighteenth century, German physician Caspar Friedrich Wolff (1733–1794) furthered Aristotle's theories of epigenesis and became one of the founders of embryology, or study of the embryo. Wolff established that an embryo's organs make a gradual appearance, and since they form in stages, they were not preformed at conception.

By the late nineteenth century, the study of cell theory gave a decided advantage to the epigenesis perspective. Scientists such as Oscar Hertwig (1849–1922) had discovered that the complexity of human development builds over time and that it is the interaction of cells from both the egg and the sperm through the process of fertilization that make a new organism. During the nineteenth century, gynecologists began to focus on the ovaries as a way to distinguish themselves from obstetricians, who took the uterus as their primary focus. The hormonal nature of the ovaries was not known at this time, but they were considered to be instrumental in secreting chemical substances that gynecologists felt regulated the female nervous system. In the nineteenth century, it was believed that the overuse of a woman's brain would cause her ovaries to shrink.

Nineteenth-century advice for women stated that the ovaries give women "all her characteristics of body in mind," and the ovaries were attributed with controlling the details of women's physical characteristics of beauty. Historically, it was also thought that sickness or personality disorders could be traced to malfunction of the ovary. Indeed, during the nineteenth century, many aliments, particularly those relating to menstruation, were said to be traced to the failure of the ovaries. There was also the notion that a woman's mind was influenced by this body part, and the so-called "psychology of the ovary" was thought to determine a woman's personality. Surgery was commonly done to upper-class women to remove their ovaries, a procedure known as ovariotomy, or "female castration," as a way to change female personality problems that revolved around sexuality and unfeminine behavior. Female castration was also a way to cure symptoms of a woman's sexual behavior and to improve a woman's morality in order to make her hardworking and easier to get along with. Today, the removal of the ovaries is a common occurrence in order to prevent ovarian cancer. However, this practice is considered controversial.

By the early twentieth century, the ovaries became more significant in medical study, particularly in the emerging field of endocrinology, or the study of hormones and how they regulate body functions. Endocrinology represented a paradigm shift in how female sexuality was considered—from a physical, biological center of sexual urges connected to an organ (the uterus) to a process related to a chemical reaction from specific hormones produced in the ovaries that regulated sexual urges. This process led from a biological definition of sex to a more technical one as hormones produced in the ovaries were increasingly studied in laboratories throughout the twentieth century.

The Ovaries and Contraception Technologies

Hormonal contraception methods of preventing pregnancy were first developed in 1960 with the introduction of the birth control pill and now include patches, vaginal rings, implants, and injections. They function by regulating the amount of progesterone and estrogen in a woman's body in order to suppress ovulation and prevent the ovaries from releasing eggs. The release of these hormones into the body through these contraception methods can also prevent pregnancy by thickening the cervical mucus which blocks the sperm from reaching the egg and also makes the lining of the uterus too thin for an egg to be implanted.

Fertility Technologies

Assisting with a women's ability to conceive a baby has undergone a vast improvement with the development of technology over the past few decades. Assisted Reproductive Technology (ART) refers to all treatments that involve surgically removing the eggs from the ovaries and combining the eggs with sperm to produce a pregnancy. These include IVF (in vitro fertilization), which removes a woman's eggs and fertilizes them outside of her body with embryos which are then transplanted into the cervix; GIFT (gamete intrafallopian transfer), a procedure that removes eggs from the ovaries, combining them with sperm, and placing unfertilized eggs and sperm into the woman's fallopian tube; and ZIFT (zygote intrafallopian transfer), wherein eggs are collected from the ovaries, fertilized, and then the zygote (fertilized egg) is surgically placed into the fallopian tubes. Other technological procedures that have improved women's chances of fertility include the process of "egg donation," in which women who are unable to produce viable eggs use those from a donor (who is usually paid). The egg donor is first stimulated with fertility drugs and then several of her eggs are retrieved. The donor's egg is fertilized with sperm and implanted into the uterus of the carrier.

Harvesting and Freezing Eggs

Harvesting eggs is a necessary process and first step for in vitro fertilization. It begins with stimulating the ovaries using hormonal fertility drugs to create ten to thirty mature eggs in the ovaries. Then a surgical procedure is performed to remove the eggs from the ovaries one at a time with a needle through the uterus. The eggs are then inspected and fertilized with sperm and implanted into the uterus to begin a pregnancy.

Women's fertility is closely related to the availability of viable eggs produced by her ovaries. This production is influenced by a variety of factors, particularly age, as women ovulate less frequently and the quality of their eggs decreases as they approach menopause. As a way to control the body's processes on the production of eggs available for fertilization and pregnancy, services have become available to harvest and freeze women's unfertilized eggs. The process involves stimulating the ovaries with fertility drugs to produce more than usual, extracting them from the woman's body, then freezing and storing them in a laboratory in order to be used at a later date for fertilization and implantation. This procedure has been advertised as a way for a woman to "set her own biological clock" and regulate her fertility to a more conducive time schedule.

A woman's age is strongly correlated to the availability of eggs that are viable for reproduction. Since a woman is born with all of her eggs, as she gets older and approaches menopause, her eggs decrease in both quality and quantity. As a woman ages beyond her mid-thirties, the quality of her eggs gradually degrades and abnormalities in the chromosomes increase, producing more miscarriages and an increased chance of birth defects such as Down syndrome.

Ovarian Cancer

Ovarian cancer is a disease in which the ovarian cells grow uncontrollably and produce a tumor on one or both ovaries. It is notoriously hard to diagnose early since there are few or mild symptoms. It is often not discovered until it is advanced and has spread to other organs, making it among the most deadly of all cancers. The use of oral contraception, pregnancy, and breast feeding has been found to reduce the risk of getting the disease. However, there is currently no standardized screening test for ovarian cancer, and it tends to most commonly affect women over the age of fifty. Women who are at very high risk of ovarian cancer may elect to have their ovaries removed to prevent the disease from occurring.

Ovarian Transplants

Though rare, there have been cases of a successful transplant of a whole ovary from one woman to another in order to have a baby. The surgery could restore normal hormone function for women going through early menopause, whether because of cancer treatments or other, unexplained causes. It also could mean that in the future, a woman with cancer could freeze an ovary, undergo chemotherapy and radiation, and have her own ovary returned later to restore her fertility.

Further Reading: Hands off Our Ovaries Web site, http://www.handsoffourovaries.com; Heinrich Von Staden. *Herophilus: The Art of Medicine in Early Alexandria*. Cambridge: Cambridge University Press, 1989; Oudshoorn, Nelly. *Beyond the Natural Body: An Archaeology of Sex Hormones*. New York: Routledge, 1994; Pinto-Correia, Clara. *The Ovary of Eve: Egg and Sperm and Preformation*. Chicago: The University of Chicago Press, 1997. www.handsoffourovaries.com

Martine Hackett

Penis

Cultural History of the Penis

The penis is the male sex organ of the higher vertebrate animal. The physiology of the human penis does not tell the complete story of its complex meanings. Biology provides the penis, but culture furnishes the ideas and customs that make the penis a symbolic organ with social and psychological meanings. The physiology of the human penis is relevant to its cultural history. The penis is designed to deposit semen, the viscous fluid containing sperm, the male reproductive cells, into the female sex organ (the vagina) for the fertilization of the female egg or eggs. The penis and the testes, the ovoid organs hanging in a sack (the scrotum) beneath the penis and the site for the production of sperm, constitute the male genitals (genitalia). The urethra, a tube connecting the bladder with the opening on the head (glans) of the penis, is the passage through which both semen and urine leave the male body. A fold of skin (the prepus, or foreskin) normally covers the glans, but the foreskin retracts when the penis is erect.

Circumcision is the medical or religious procedure in which the foreskin is cut away. Erection of the penis occurs when the chemicals released during sexual arousal cause an increase of blood to the penis (about ten times the normal volume), where physiological mechanisms keep the blood in the soft, spongy tissue (corpus spongiosum) so that the penis enlarges and stiffens, until some event, such as orgasm (the physiological event, usually accompanying the ejaculation of the semen from the penis) or an external distraction, permits the blood to leave the penis and for the penis to return to its normal, soft (flaccid) state. The size of the human penis varies, but several surveys suggest that most men have a penis between five and seven inches long (erect, average six to six-and-one half inches) and between four-and-one-half and five-and-one-half inches in circumference. A common form of male contraception is the placing of a condom (sometimes called a sheath) on the penis to catch and contain the semen during ejaculation.

The physiology of the penis, in particular the necessity of an erection for the penetration of the male organ into the female vagina, has much to do with cultural attitudes toward the penis. The penis comes to stand for male power, embodying male dominance, aggressiveness, agency, and creativity. Males sometimes use the penis to dominate women and other men; male anal penetration of another male is seen as feminizing the receiving male. The phallus, another name for the penis, is the term used by scholars to describe this powerful, symbolic penis. In psychoanalytic thought, one controversial claim is that girls come to envy the penis and its power. Yet another psychoanalytic approach sees much of male culture as compensation for the "womb envy" men have of women, who have the power to produce new life. This may lead to men's exaggerated view of the power and creativity of the penis.

Male initiation rituals at puberty, the age at which the adults are ready to signal the transformation of a boy into a man, often put the penis at the center of the ceremonies, as when the initiation involves circumcision. Even in the absence of formal puberty rites, boys see the ability to ejaculate semen as a sign of their transition from boy to man, equivalent to the onset of menstruation in women.

The Penis as Phallus in Pagan and Non-Western Cultures

Non-Western cultures and the classical cultures (Greece and Rome) antecedent to the emergence of a Christian-dominated West view the penis as a creative source of power and life. Many cultural creation myths look to the god's penis or masturbation by the gods or copulation by the gods as the source of humanity and fertility. In some cultures, penis icons and images represent this creativity and fertility. The Tagata Jinja (Shinto) shrine near Nagoya, Japan, is famous for the tradition of carving a large penis from a cypress log each spring and parading it as part of the spring fertility rite.

Ancient Egyptians saw the penis as sacred and life-creating and, like the Hebrews, circumcised their boys. The ancient Greeks did not believe in circumcision, and historians point to visual and written evidence for the view that Greeks valued small, uncircumcised penises. The *kouri*, the statues of nude young men found throughout the ancient Greek world, seem to capture the Greek ideal. Large penises were thought crude, the property of lesser races and people. Exercising in the gymnasium and other manly pursuits in ancient Greece were done in the nude, but in these public settings the penis was infibulated, the foreskin pulled forward over the glans and held into place with string or with a circular pin called the *fibula*. Greek men commonly greeted each other by touching the other's testicles, and the English words "testify" and "testimony" come from the custom of swearing truth by grasping one's own testicles. The erect penis symbolized Athenian power to the citizens of that state, and archaeologists find *phalloi* (penis stone replicas) throughout the ancient Greek world. The Greek idealization of the male body meant that pederasty—the anal penetration of a boy by his mentor's penis—was an accepted rite of passage into manhood, and the Greeks attributed power and manliness in this exchange of semen. The anal penetration of an adult male, however, had quite different meanings, feminizing the penetrated man.

In Rome, penis replicas (the *fascinum*) were also common, but the Romans favored large penises and took special delight in Priapus, a small god usually depicted with a penis half his size and eternally erect (hence the present medical term for a persistent, unwanted erection—"priapism"). The Romans rejected the Greek view that the virtues of manliness could be transmitted from a man to a boy through semen; anal penetration always meant feminization to the Romans, but older Roman men would give lockets, *bulla*, to favored young men, and in these lockets were small fascinum. The large, erect penis represented a strong, aggressive, martial masculinity to the Romans, and historians note that the Latin word for bullet, "*glans*," makes clear the association of these projectiles (hurled with slings) with rape—the penetration of the enemy's body.

The Penis Redefined by Christianity

Historians agree that a turning point in the cultural history of the penis in Western civilization was St. Augustine's idea of original sin and the corrupting influence of sexual desire. For Augustine, the penis, erections, semen, and desire were forces drawing men away from God. By the thirteenth century, the penis largely disappeared from Western painting and sculpture. The only exception was the penis of baby Jesus, which served to prove he was both divine and human, having a human body in all its features. Jesus's foreskin even became a sought-after religious relic. On the other hand, the demonization of the penis by Christianity was very clear during various witchcraft epidemics from the fourteenth through the seventeenth centuries, when a common confession by an accused witch admitted to her fornication with Satan or with one of his devils, and some witchcraft accounts go into details about Satan's penis.

Leonardo da Vinci (1452–1519) and other Renaissance artists were able to paint and sculpt the penis in public art by the sixteenth century, but still only within limits set by the church. Da Vinci's early anatomical examination of the penis and the testicles led within a generation to other scientific studies, notably those by Andreas Vesalius (1514–1564), the Belgian whose 1543 publication of *De humani corpus fabrica* (*On the Fabric of the Human Body*) set a new standard for understanding human anatomy. All through this period, the actual functioning of the penis and testicles, and the role of sperm in reproduction, were not well-understood. In fact, it was not until Antoni van Leeuwenhoek (1632–1723) examined semen under a microscope that the sperm was identified. This microscopic look reinforced the preformationism theory that the sperm contains a preformed human. It was not until the 1870s that scientists finally understood the fertilization of the female egg by the sperm.

Masturbation

An important part of the cultural history of the penis in Western societies is the history of ideas about male masturbation (rubbing the penis to produce an orgasm and ejaculation unrelated to reproduction, sometimes called "onanism"). When anthropologists have mentioned the attitudes toward masturbation in the societies they study, usually they discover that people other than Westerners have a casual, matter-of-fact attitude toward the practice. But in the eighteenth and nineteenth centuries in the United States and England, the Augustinian heritage in Christianity meant that masturbation was viewed a sin to be avoided. Even physicians dispensing advice to parents and young men in child-rearing and character-training manuals in the eighteenth and nineteenth centuries saw masturbation as an unhealthy habit that should be avoided. The scientific reason for this often was the view that semen contained the essence of the vitality of a young man, and that he should release his semen only for the socially acceptable purposes of procreation. In this view, masturbation would lead to a series of physical and social ills in the boy, draining him of his "vital fluids" and obsessing him to the detriment of all of his social relations and obligations.

The combination of religious and medical condemnations of masturbation, well into the twentieth century, served to make masturbation the source of great anxiety and guilt for boys and young men. We know from many sources that the prevalence of male masturbation in the West probably has not changed much over the centuries, but cultural attitudes about masturbation have meant that many boys in the West grow up thinking of sex and their penises as something "dirty" and secret. There are physiological and possibly sociobiological reasons why male human sexuality is centered on the penis, but cultural attitudes toward the penis and masturbation in the West have made these activities a cause for secrecy and shame.

Male Anxieties

Cultures have differed over time and space in their view of the ideal penis, especially its shape and size. A man with a penis that does not fit the cultural ideal experiences some level of psychological stress. The ancient Greek with a large penis and the ancient Roman with a small penis, we assume, would face such anxieties. It seems that men in Western societies are quite anxious about their penises.

Western societies from the Industrial Revolution to the present have valued large size. The American proverbial phrase, "bigger is better," applies to a range of judgments and artifacts in the West, so it is not surprising that white Americans and Europeans would consider large penises (in length and girth) to be "better." Some historians have traced this concern with size to the white European's colonial encounter with black Africans. Although actual measurements of the length and girth of the penises of black men and white men (as uncertain as these racial categories are) are scarce and self-reports on penis length and girth are notoriously unreliable, there is still no evidence that the average size of a black man's penis is larger than that of a white man. Yet the belief flourished from the fifteenth century on, reinforced by the white view that the black African is closer to animals than to humans. The biblical story of Ham also contributed to this myth. White American men, especially in the South, reacted to the enslaved black and even to the free black man in ways that historians have interpreted as white fear of the well-endowed black man, picturing him as an animal-like and in pursuit of sex with white women.

Information on penis size is very suspect. Most of the surveys rely on self-reporting, and even with explicit instructions on how to measure the penis's length and girth, these reports seem to most experts to be unreliable.

The very few studies that report data from actual measurements taken by trained staff show that most erect penises lie between 5.5 inches (14 centimeters) and 6.3 inches (16 centimeters). But the actual size of men's penises is less important than the findings of surveys that many men are unhappy with the size of their penises and, if they could change anything about their bodies, they would want larger penises.

It is not just size that gives Western men anxieties. The penis is not the phallus. As a symbol of power and potency, the phallus rarely fails. Real penises, however, are not always under the control of the men to whom they are attached. Sometimes the penis becomes erect when the owner does not want an erection and sometimes it fails to become erect when the owner wants an erection. In Western cultures, people have said that the penis has "a mind of its own," doing as it pleases, which reflects the odd relationship men have with their penises. The man experiences the penis sometimes as part of his body and sometimes as a thing apart. Several jokes refer to the man's thinking with the head of his penis rather than with the head on his shoulders, and there are dozens of slang words for the penis in many languages.

Jokes and other folklore provide rich evidence of the ways in which men express their anxieties about the penis's size and performance. Sigmund Freud (1856–1939) and other psychoanalytic writers have made the penis (the phallus) a central symbol of masculine power in patriarchal societies. Freud saw folklore, dreams, jokes, and fantasies as expression of anxieties and thoughts forbidden to the conscious thought. Thus, a man will not worry openly about his penis, but he will tell and laugh at jokes and stories about penis size and performance.

One way a penis can fail a man is by remaining soft when he wants it erect. Impotence (the inability of the man to achieve and sustain an erection) is an old affliction; ancient societies had folk cures for impotence, and the machine age brought mechanical solutions. The recent obsession in American culture (and elsewhere) with medical approaches to curing impotence has provided much evidence for historians and other scholars of the symbolic importance of the erect penis as a sign of potent masculinity in American society. Impotence has many medical and psychological causes, but the promise of strong erections by taking pills (Viagra and related drugs) fits well the American desire for quick and easy solutions through drugs.

The Invisible and Visible Penis

Upright posture in humans made the penis visible, while the female genitals remained invisible. Culture has more often found ways to cover the penis than leave it open to view, adding to its mystery and power. Clothing and body adornment have drawn attention to the penis at different times and in different ways. The extended penis sheath worn by indigenous people in Africa, South America, and the Southwest Pacific enhances the display of dominance represented by the erect penis, and even when the penis is covered by clothes in Western societies, its presence may be signaled. The codpiece was an item of clothing (popular in fifteenth- and sixteenth-century Europe and England) meant to conceal the genitals visible through tight stockings, but the effect (not unintended) was to draw attention to that part of the male body; this paradox of highly visible invisibility led to elaborate decorations, padding, and other enhancements of the genital region. Male dancers of the classical ballet, also in tights, enhance the size of their genitals with padding. More recently, tight clothing has been used to display the penis without exposing it. Tight pants (especially blue jeans), swimsuits, and underwear in print and electronic advertisements often make clearly visible the outline of the concealed genitals.

Despite the strong religious and social objections to the visible penis in the late nineteenth and twentieth centuries, the history of Western art and photography (invented in 1839) shows the gradual emergence of the visible penis. Art photography of male nudes, appearing as early as the 1850s, made the penis visible, but largely under the guise of male physique studies and imitations of classical painting and sculpture. Photographers in the West actually photographed full frontal nudity earlier, and far more often, than did their painting colleagues. The "animal locomotion" photographs by Eadweard Muybridge (1830–1904) in the 1880s showed stop-action sequences of photographs of nude men walking, running, and engaging in other natural, everyday movements. The rise of a cult of muscular male beauty near the end of the nineteenth century doubtless stemmed from a perceived crisis in masculinity in the United States and Europe, and strongman Eugene Sandow (1867–1925) represented the new ideal of the highly muscular male. While Sandow's genitals were covered by the "fig leaf" or posing strap in photos from the early twentieth century, photographers of male nudes began showing male genitals by the 1920s. The visible penis also matched the emergence of nudism and "physical culture" in the United States and Europe, especially Germany, in the 1920s and 1930. Beginning in the 1930s, gay male photographers produced art photographs and commercial "physique" photographs in magazines catering to a male homosexual audience, and these images moved from the posing-pouch-covered genitals in the 1950s and 1960s to fully nude photography by the 1970s. In the late 1980s, women art photographers were also producing male nudes exposing the penis. Meanwhile, mainstream film and then cable television saw milestones in the exposure of male genitals.

Not generally visible are the practices of tattooing and piercing the penis and scrotum. The earliest form of male genital piercing was the "Prince Albert," a ring piercing the urethra and glans and reportedly named for such a piercing Queen Victoria's husband used to keep his penis tucked away. By the late twentieth century, many other sorts of male genital piercings also became popular.

Some performance artists in the late twentieth century began creating pieces using the male nude body and the exposed penis. The visibility of the penis as performance art took a new form in 1997, when two Australian men created a show, *Puppetry of the Penis*, in which they stretch, bend, and fold their penises, scrota, and testes into various shapes resembling other objects. They toured many countries, including the United States, from 1998 through 2007. *See also* Genitals: Genital Piercing; Muscles: Cultural History; and Penis: Penis Envy.

Further Reading: Bettelheim, Bruno. *Symbolic Wounds: Puberty Rites and the Envious Male*. New York: The Free Press, 1954; Bordo, Susan. *The Male Body: A New*

Look at Men in Public and Private. New York: Farrar, Straus and Giroux, 1999; Cohen, Joseph. *The Penis Book*. Cologne, Germany: Konemann Verlagsgesellschaft, 1999; Dutton, Kenneth R. *The Perfectible Body: The Western Ideal of Male Physical Development*. New York: Continuum, 1995; Friedman, David M. *A Mind of Its Own: A Cultural History of the Penis*. New York: Free Press, 2001; Leddick, David. *The Male Nude*. Cologne, Germany: Taschen, 1998; Lehman, Peter. *Running Scared: Masculinity and the Representation of the Male Body*. Philadelphia: Temple University Press, 1993; McLaren, Angus. *Impotence: A Cultural History*. Chicago: University of Chicago Press, 2007; Mechling, Jay. "The Folklore of Mother-Raised Boys." In Simon J. Bronner, ed., *Manly Traditions: The Folk Roots of American Masculinities*. Bloomington: Indiana University Press, 2005, pp. 211–227; Paley, Maggie. *The Book of the Penis*. New York: Grove Press, 1999.

Jay Mechling

Male Circumcision

Male circumcision, or the removal of part or all of the two layers of foreskin and possibly the protective skin of the penis known as the frenulum, is the most common surgical procedure in the United States. Though rates of circumcision among boys have decreased in the United States since the 1970s, there are still well over 1.25 million boys circumcised annually in the United States alone. Circumcision is a centuries-old practice, with cave drawings dating back to the Paleolithic period depicting circumcised men and the knives and stones used to circumcise them.

Circumcision is currently a highly debated topic with strong opponents arguing against its necessity. Males ranging in age from newborns to adult men are circumcised for a number of reasons; these have changed historically,

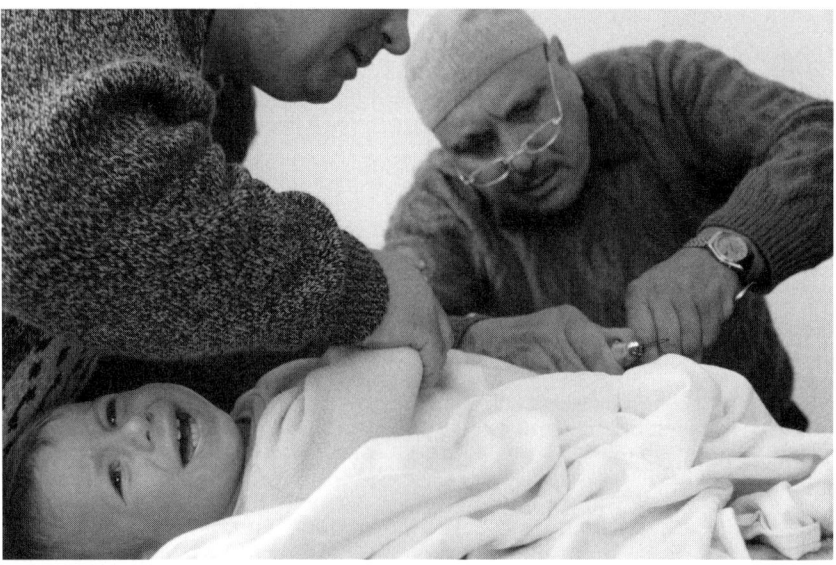

Bulgarian boy from the Pomaks ethnic minority reacts as a doctor performs a circumcision on him during a mass circumcision ritual in the village of Ribnovo, 2006. (AP Photo/Petar Petrov)

and range in nature from religious, cultural, hygienic, aesthetic, psychological, therapeutic, to medical.

Physiology

The penis is comprised of the shaft, the meatus or the opening at the tip of the penis (referred to as the glans or the head of the penis), the frenum, the foreskin, and the corpus spongiosum. While in a non-erect state, the only visible areas of the penis are the shaft and the foreskin, along with the small opening at the top of the foreskin. The foreskin is designed to protect the glans. There are two layers of foreskin, an outer layer and an inner layer. The foreskin covers the sulcus, glans, frenum, and urinary meatus. As the length of the penis increases during erection, the foreskin gets stretched as the glans moves outside the foreskin opening and is exposed. During circumcision, either all or part of the foreskin is removed along with part of the frenulum, the skin tissue located under the glans.

Methods

Circumcision can take place in a number of different ways, depending on the age of the patient and the cultural circumstances in which the procedure occurs. In the United Sates, infantile circumcision takes place a few days after birth. The infant is restrained and cleansed. General anesthesia is not used during infantile circumcision as it has been deemed too risky for infants; however, sometimes, local anesthesia is used. During adult male circumcision in the United Sates, anesthesia is usually used.

During surgical circumcision, a clamp is placed around the glans or the head to protect it from any damage and to separate the foreskin from the glans. The foreskin around the glans is then sliced and cut around the glans to separate and to remove the skin. This is done with either surgical scissors or a scalpel. Pressure is used to control the bleeding. With infants, stitches are not needed and the surgery takes a few minutes.

In the United States, circumcision can also take place using a Gomco clamp. This clamp is used to squeeze the foreskin off the tip of the penis, reducing the likelihood of bleeding. The foreskin is first separated and pulled up, stretching it over the glans. The foreskin is slit so that the bell-shaped component of the Gomco clamp can be fitted in between the glans and the foreskin. This is partially done to protect the glans from any damage. The clamp is then fitted and tightened, squeezing the foreskin to prevent blood flow. After several minutes, the foreskin is sliced away with a scalpel and the clamp is removed.

Another similar procedure to that of the Gomco clamp is the Plastibell. The Plastibell is a plastic, bell-shaped object with a small hole in it used for urination. In this procedure, the foreskin is separated from the glans. The Plastibell is placed in between the glans and the foreskin. The foreskin is tightened with a string to prevent blood flow. After blood flow is reduced, the foreskin pulled up and the foreskin pulled above the cap is cut away, leaving the Plastibell in place and the foreskin at the base of the glans in place. Several days after the procedure, the plastic bell will fall off. Most infantile circumcisions take place using the squeezing method of the Plastibell or the Gomco clamp, while most adult circumcisions use the surgical method. Many argue that the extent of pain involved in infantile circumcision is minimal, while the pain of adult circumcision is significantly greater.

There are complications that can occur during the procedure, which range from accidental removal of parts of the actual penis to too much removal of the foreskin (which causes erections to be painful). However, most of the problems that occur during male circumcision take place after the surgery and result from infections. Complications from infections can result in a range of negative outcomes, from removal of the penis to death. The controversies that have resulted from such complications will be discussed further in this entry.

Circumcisions in the United States began in the mid-1800s and increased substantially over the course of the next hundred years. There are a number of reasons why male circumcisions take place, both medical and cultural. Circumcision has been believed to be more hygienic, and to prevent disease. Circumcision was once prescribed by doctors to discourage masturbation. Religious and moral sentiments have recommended against masturbation. Not only was it considered a sin in many religions, it was also believed to cause a number of medical ills, from asthma to kidney disease, headaches, epilepsy, alcoholism, and even insanity.

Medical Reasons

In the United Sates, circumcision is seen as a medical procedure. Circumcision is believed to remedy malformed foreskin. Foreskin can be either too long or too short. Foreskin that is too long will not properly allow the glans to go through the opening during erection. It will also prevent proper ejaculation and urination. A foreskin that is too short can cause painful erections as the skin will tighten during erections. This condition is called phimosis, occurring when the foreskin is tight and unable to retract. A circumcision on an infant or an adult is often prescribed to remedy these issues.

Medical support of circumcision on medically normal penises is based on hygiene. Some have hypothesized that circumcision developed as a way of keeping clean in the desert, where there was much sand and little water for bathing. Contemporary supporters of circumcision believe that the practice reduces the chance of infections within the foreskin. Supporters for circumcision also argue that it has a positive impact on a number of other health issues, such as cancer and the transmission of diseases such as AIDS.

Those who advocate for circumcision maintain that it reduces the chances of different forms of penile, prostate, and cervical cancer. Supporters argue that there are lower instances of penile and prostate cancer among men who are circumcised; however, the reasons have yet to be explained. Some also believe that circumcision will reduce the chances of cervical cancer in women, by lessening the chances of intercourse-based transmission of the human papillomavirus (HPV), which is linked to cervical cancer.

AIDS and Other Sexually Transmitted Diseases

Recently, circumcision has been championed as a public health tool in the global fight against AIDS. A recent study conducted by researchers in Uganda, Kenya, and South Africa found that within the sexually active heterosexual population, male circumcision reduced the chances for HIV transmission by almost 60 percent. These findings have influenced public health advocates, including the World Health Organization, to promote male circumcision.

Some researchers and government and health officials are now recommending circumcision as another preventative measure along with condom use and abstinence. Some public health officials, however, worry that recommending circumcision as a way to decrease HIV transmission may influence people who are circumcised to feel immune and engage in unsafe sexual practices. Critics argue that circumcision may be unnecessary or unhelpful, with a number of complications, as described above.

Cultural Circumcision

Circumcision is an ancient practice that is highly influenced by culture and religion. In some of the major world religions, including Judaism and Islam, circumcision is religiously prescribed. Hindus and Sikhs, most of whom live in India, do not practice circumcision, and in Christianity circumcision is sometimes prohibited, although some Christian groups require it.

Judaism

Circumcision is an important part of the Jewish tradition. The Torah, the Jewish religious holy text, dictates that Jewish boys be circumcised. God commanded Abraham and all of his male decedents to be circumcised, and as a result, all Jewish males are required to be circumcised to symbolize God's covenant with his followers. During a ceremony called a *bris milah*, commonly called a bris, a mohel performs the circumcision. This ceremony usually takes place eight days after the child's birth. Originally, only the tip of the foreskin was to be removed. However, full removal of the foreskin and often the frenulum, a more radical circumcision, is seen as a means to prevent Jewish men from pulling their foreskin forward to assimilate with other groups.

The spilling of the blood from the penis is an extremely important aspect of circumcision. If a boy is born without foreskin, the child is cut in order to spill blood. The same extraction of blood takes place for an adult male converting to Judaism who has already had a circumcision.

Though rare, some Hasidic sects engage in a practice called *metzizah*, where the mohel sucks the blood from the wound to clean the wound and encourage healing. This is a very controversial practice, one of the reasons being the possible spread of disease from the mohel to the child. As a result, often the mohel's mouth is sanitized or the mohel may use a glass tube and place it in between the wound and his mouth.

Islam

Though an important part of Islam, the Koran, the Muslim scripture, makes no mention of circumcision. However, many followers of Islam profess that the prophet Mohammed promoted circumcision. One reason often cited for circumcision is to keep the genitalia clean, and cleanliness is an important part of Islam. It is a *sunnah* act, or an act that follows in the tradition of Mohammed, and his followers. The time of circumcision can vary based on the age, class, and even geographic location, but circumcision usually occurs by puberty. If a boy is born in a hospital, the circumcision usually takes place shortly after birth; otherwise, children can be circumcised at around six years of age, and many boys are circumcised at the start of puberty. Though rare, there are some Muslim groups that permit circumcision during adulthood, usually for non-circumcised converts.

Circumcision that takes place during youth or at the start of puberty usually occurs during a religious ceremony preceded by fasting. No general or local anesthesia is used. During the procedure, the foreskin is pulled over the glans and removed, often leaving the inner foreskin.

Christianity

For the most part, Christianity does not require circumcision of its followers. In fact, a number of books of the Bible recommend against it, seeing it as unnecessary. Yet, some Orthodox and African Christian groups do require circumcision. These circumcisions take place at any point between birth and puberty. The Catholic Church proscribes circumcision for Christians, and as a result, many Catholic men born outside the United States are not circumcised.

Other Groups and Cultures

In some parts of Asia, for example, South Korea, circumcision is popular, a phenomenon often attributed to Western influence after the Korean War. The same is also the case in Japan, where circumcision has been gaining in popularity since World War II; again, this has been attributed to Western influence. Circumcision has been particularly popular among the younger generations.

Circumcision and other forms of penile modification are common parts of initiation rituals or rites of passage in many cultures. Circumcision can vary from a small cut on the foreskin (usually practiced by Pacific Islanders) to the removal of the foreskin all together, to a practice known as *salkh*. Salkh is not common, but is practiced by small Muslim tribes in Yemen. In a ceremony conducted before marriage, the bridegroom has the skin of the penis going up to the lower abdomen and down to the scrotum removed. This is performed without anesthesia, and the initiate is required to endure the procedure without expressing pain or showing discomfort. Failure means that the ceremony is called off, and the initiate is prevented from marrying.

Among a number of African tribes, circumcision is a rite of passage and depending on the tribe, can take place at various points of a young man's life.

For many of these different groups, male circumcision is also performed for aesthetic purposes. Similar to arguments in the United States which support circumcision, advocates for circumcision argue that the circumcised penis looks "normal" and that most sons should look like their fathers or the other men and boys in their communities. In addition, they also argue that many women expect circumcision and may not be attracted to uncircumcised men.

Subincision is an extreme form of male genital modification practiced by the Aboriginal peoples of Australia. This is normally performed on teenagers and consists of slicing the underside of the penis from the urethra to the testicles. The procedure starts with a meatotomy, or slicing open the head of the penis to the urethra. This is often performed with a small, sharp rock or knife and no anesthesia. After the penis is sliced open, small thorns are inserted into the cut. The skin heals around the thorns while the urethra is still exposed. Men are able to still maintain an erection and urinate; some argue that sex becomes more pleasurable after such a modification.

Controversy

There is currently great debate surrounding the necessity for routine circumcision on baby boys. Many of the reasons cited for circumcision have been debunked, such as preventing or reducing masturbation, reducing rates of cancer, and promoting hygiene. Critics point out that a number of complications can occur as a result of circumcision. The patient can experience blood loss or hemorrhaging and infection. If not properly treated, the infection can lead to removal of the penis or even death. Death is rare: researchers estimate that one in every 500,000 circumcisions result in death. However, the practitioner can make an error and accidentally slice part of the glans; there have also been cases where the penis has been damaged or too much of it has been accidentally removed. The possibility of too much skin being removed also occurs. When this happens, erections can be painful. Critics of male circumcision have also argued that the penis loses its sensitivity after circumcision, because the glans of a circumcised penis, normally protected by the foreskin, is exposed. The nerves in the glans become less sensitive and may reduce the pleasure experienced during sexual activity.

Importantly, patient consent is a major issue in circumcision debates in the United States because the vast majority of those being circumcised are infants. Anti-circumcision activists argue that circumcision is a decision made by parents and doctors rather than the individual being operated on, and suggest that it is an aesthetic surgery that serves only to mutilate the penis. Activists argue that circumcision has no real purpose beyond simply satisfying cultural norms and values. Some see male circumcision as akin to female circumcision, in which the labia and/or clitoris are removed during a highly controversial religious and cultural rite of passage. Most societies that practice female circumcision also practice male circumcision. Both male and female circumcisions are justified for aesthetic, moral, hygienic, cultural, or religious reasons. *See also* Genitals: Female Genital Cutting; and Penis: Subincision.

Further Reading: Denniston, George, Frederick Mansfield Hodges, and Marilyn Fayre Milos. *Male and Female Circumcision: Medical, Legal, and Ethical Considerations in*

Pediatric Practice. New York: Springer, 1999; Favazza, Armando R. *Bodies Under Siege: Self-Mutilation in Culture and Psychiatry.* Baltimore, MD: The John Hopkins University Press, 1987; Glucklich, Ariel. *Sacred Pain: Hurting the Body for the Sake of the Soul.* New York: Oxford University Press, 2001; Wallerstein, Edward. *Circumcision: An American Health Fallacy.* New York: Springer, 1980.

Angelique C. Harris

Penis Envy

Penis envy is a psychoanalytic concept that was introduced by Sigmund Freud (1856–1939) to explain female psychosexual development in his 1908 article "On the Sexual Theories of Children." A driving force in Western perceptions of female sexuality, it describes a young girl's discovery that she does not have a penis, the subsequent effects of which are crucial to the development of her femininity and heterosexuality. According to Freud's theory of psychosexual development, penis envy is meant to occur during the third, or phallic, phase of development. It is during this stage that a young child's libido (broadly defined by Freud as the primary motivating force within the mind) becomes fixated on the phallus. This fixation engenders an Oedipal crisis in both genders; however, the difference in genital anatomy results in divergent, though respectively uniform, responses in girls and boys.

A brief overview of Freud's theory of psychosexual development helps to contextualize and elucidate the concept of penis envy. The theory supposes the presence of a libido defined as an inherent psychic energy that is transferred to and from different objects for the purposes of personal development and, eventually, individuation. Freud assumes the existence of an original libido of the ego (ego-libido) that is transferred, or redistributed, to other objects at the time of pivotal experiences during development. These pivotal experiences mark transitions into and between several phases (oral, anal, phallic, latency, and genital); a person's successful entry into and completion of each of these phases defines their trajectory toward normal sexuality. Penis envy is understood to occur during the phallic phase, anywhere from thirty-six months to seventy-two months of age, when the libidinal energy shifts from the anal region to the genital region. It is not so much the discovery of the penis, but rather the young girl's recognition that she does not have one, that catalyzes her experience of what Freud calls the Electra complex.

According to Freud, a child develops a sexual impulse toward her or his mother shortly following the libidinal shift to the penis. However, the female child will realize, thereafter, that she is not physically equipped to have a sexual relationship with her mother and will, in turn, develop a desire for the penis, or rather the phallus and the power that it represents. She will blame her mother for what she perceives as a lack, and her desire will shift its attention to her father. Fearing, however, that these desires will incur punishment, she defends herself by displacing her father as the object of her desire with men in general.

Though not explicitly, Freud more fully developed the theory of penis envy in 1914 in his pivotal essay "On Narcissism." In the essay, his analysis of the title topic is analogous to and interwoven with other instances in which sexual desires represent a redirection of the ego-libido toward other objects, body parts, people, or concepts. If, according to Freud, narcissism is a "perversion" that is wholly absorbing, we can understand it as a libidinal fixation that is, in this case, directed inward, analogous to a girl's fixation on the phallus during the phallic phase of her psychosexual development (1914). In the essay, Freud discusses the relationship between the ego and external objects and, consequently, draws a distinction between the "ego-libido" and the "object-libido." While narcissism is indicative of a synchronicity between the ego-libido and the object-libido, the emotional investment in the phallus, the "phallic-cathexis" for men can result in narcissism but, for women, is a directing of the libido outward. Narcissism is, at least to some degree, presented as a normal instinct within male development and, in contrast, an occasion of vanity in woman.

Freud implies that for women, for whom penis envy is the counterpart to castration anxiety, the emotional response to and the feelings induced by the penis do not lead to narcissism, precisely because they do not see themselves reflected in the phallus. Rather, women would, within this framework, interpret their vagina as the lack of a penis and interpret the penis as an object onto which they can place their desires and resolve their lack. In 1996, psychoanalyst Gerald Schoenwolf developed the concept of gender narcissism to build on Freud's theories of penis envy and the castration complex. He suggests that an overemphasis or overperception of gender and gender differences in childhood can lead to either devaluation or an overvaluation of one's gender in later life. Indeed, the narrow interpretation of a child's experience of his or her genitalia may itself contribute to a limited view of what constitutes normal sexuality and sexual identity.

Freud's assumptions about gender have aroused censure and rebuttal from feminist critics. Feminists object to the idea that female sexuality is constituted by a lack and, that, psychologically, women are deficient men. Further, the theory of penis envy specifically conceptualizes the significance of the phallus in favor of men. The theory of penis envy reifies the phallus, equating the penis with power. In this framework, power is defined as something distinctively un-female. According to this logic, a woman's first recognition that she doesn't have a penis will automatically translate to her as a lack and a signal that, in not having a penis, she does not have power. In addition, this theory treats the vagina as an emptiness rather than as something present and corporeal. Furthermore, a girl's experience with her own genitalia, to the extent that it is acknowledged at all, is defined in terms of a crisis. Her nearest resolution of this crisis of powerlessness, according to this theory, will be limited to her romantic relationships with men.

Feminists have suggested that Freud's theory assumes the sociocultural overvaluation of the male over the female. Within the scope of Freud's upper-class and morally conservative surroundings, penis envy was a logical organizer of gender classification and, therefore, fundamental to the explanation of female sexuality. For some contemporary psychoanalysts, a normal femininity has its own developmental line and is not derived from a primary, disappointed masculinity and penis envy (The American Psychoanalytic Association, 1990: 140). Such observations (made by contemporary psychoanalysts such as Harold

Blum, William Grossman, Walter Stewart, Eleanor Galenson, and Herman Roiphe) challenge traditional ideas about penis envy with a more extensive range of events and instances of self-discovery that influence later development. Some psychoanalysts have argued that the pathology implied by the term "penis envy" has more relevance in relation to disturbances of "normal" development. For example, in the instance that the mother strongly projects a sense of weakness or inferiority to the father, a girl may interpret her own genitals to define a deficiency or weakness and may wish for a penis as a substitute object assumed to be more gratifying. Persistent or exaggerated penis envy, that which greatly surpasses the transient recognition and subsequent processing of a distinction in genitalia, could be seen as an indicator of other problems or crises with regard to normative development.

A comprehensive analysis of Freud's theory of penis envy includes its scholarly and popular criticisms. Karen Horney, Alfred Adler, Melanie Klein, Erik Erikson, Jean Piaget, Luce Irigaray, Julia Kristeva, Hélène Cixous, Jacques Lacan, Jacques Derrida, Juliet Mitchell, Michel Foucault, Gilles Deleuze, and Félix Guattari are among the philosophers, feminists, and psychoanalysts who offer a range of criticisms of and elaborations upon Freudian psychoanalytic theory.

Further Reading: Brenner, Charles, M.D. *An Elementary Textbook of Psychoanalysis.* New York: Doubleday, 1974; Freud, Sigmund. "On Narcissism: An Introduction." *The Standard Edition of the Complete Psychological Works of Sigmund Freud*, vol. XIV (1914–1916), 67–102; Moore, Burness E., and Bernard D. Fine, eds. *Psychoanalytic Terms & Concepts.* New Haven, CT: The American Psychoanalytic Association and Yale University Press, 1990; Sandler, Joseph, Ethel Spector Person, and Peter Fonagy, eds. *Freud's "On Narcissism:" An Introduction.* New Haven, CT: Yale University Press, 1991; Schoenwolf, Gerald. "Gender Narcissism and its Manifestations." National Association for Research and Therapy of Homosexuality, http://www.narth.com/docs/1996papers/schoenwolf.html/.

Alana Welch

Subincision

Subincision is the practice of slicing open the underside of the penis from the urethral opening of the glans down toward the scrotum. The most notable and most widely documented subincision rituals are those of the Aboriginal tribes scattered across Australia, although localized examples have also been noted among indigenous populations elsewhere, including in Hawaii, Kenya, Fiji, Tonga, and Brazil. The origins of subincision for the indigenous Australians in particular are lost to history, and there is contention among anthropologists as to the practical purpose and ritualistic symbolism of the procedure and the mythology behind it.

In Australia, the area in which subincision has been practiced covers most of the modern-day states of Western Australia, South Australia, and the Northern Territory, although the practice seems not to have been as common in the east. Although the indigenous populations of the continent are not a single, homogenous group, the various communities do share much in the way of culture, including the principal features of the ritual surrounding the subincision. Among those tribes that embrace it, subincision has been a regular and routine feature of male adolescence.

The example of the Arunta tribe is typical. In order to reach the full status of adulthood, every male member of the tribe must undergo a series of four initiation ceremonies, of which the acquisition of a subincised penis is only one. The first rite of manhood is performed between the ages of ten and twelve, before the onset of puberty, and involves the young boy being painted, thrown repeatedly into the air, and then pierced through the nasal septum. The second takes place some years later, during which the boy is abducted by his older male relatives and, often struggling, carried off to a designated ceremonial site. This abduction marks the beginning of a seven-day period of sacred performances focusing on the boy's future marriage, and at its culmination, he is ritualistically circumcised, ostensibly to officially symbolize his masculinity and coming of age.

Once the circumcision wounds have healed some weeks later, the third rite, known as the *arilta*, may be undertaken. With no women present, it begins with ceremonial performances enacting the mythologies of the tribe's distant past. At an appointed moment, the novice is told to lie flat on his back, and is given no warning as to what awaits him. The unwitting adolescent is held aloft by two initiated elders, his penis is grabbed by a third man, and a forth approaches him with a sharpened flint blade. Using this tool, the young man's urethra is opened up with a series of cuts, beginning at the tip of the penis and extending some length down the underside of the shaft toward its base. The Arunta initially create an incision that is approximately an inch in length, though this varies from tribe to tribe.

The blood from the wound is ceremoniously collected to be poured onto a sanctified fire. The initiate proceeds to urinate on the embers of this same fire and allows the resultant steam to rise over the incision, a practice that apparently is intended to ease the pain. While he is recovering, laying flat on his back, other members of the tribe who have gathered to witness the initiation may voluntarily enlarge their own subincisions by undergoing further operations. This stage of the initiation concludes with a series of symbolic acts intended to illustrate the young man's final separation from the care of his mother. After a three-day period of enforced silence, he is considered fully initiated, though not yet a fully developed man. This finally occurs after the forth rite, another set of spiritual reenactments.

The specific practice of subincision is clearly undertaken for ritualistic purposes in this context insofar as it forms a crucial part of every boy's journey from adolescence into adulthood. Nevertheless, the Aborigines themselves never ascribe any specific symbolism or meaning to subincision beyond its status as a component of their traditional rituals. As it was practiced in the past, so it must be done today. This has baffled and frustrated anthropologists and ethnographers, who have long speculated as to the original intentions; some theories are more credible than others.

One of the earliest of these was the suggestion that subincision might have been useful as a mode of reducing population growth in areas where food and water were scarce. This theory postulated that as subincision moves the urethral opening some way down the penile shaft, it might cause the man to ejaculate outside of his partner's vagina during intercourse and consequently function as a form of contraception. This is improbable, as it misses the fact that a subincised penis could still ejaculate with enough force to propel seminal fluid into the vagina, as well as the observation that the geographic distribution of subincision rituals throughout Australia does not coincide with areas which face particular pressure on resources.

Another theory as to the basis of Aboriginal subincision suggests a totemic and overtly sexual purpose. Some scholars have theorized that subincising the penis might have begun in emulation of the bifurcated penis of the kangaroo, a powerfully symbolic animal in Aboriginal culture. The kangaroo is particularly sexually aggressive, engaging in intercourse for several hours at a time, and given the ritual framework of subincision as a path to manhood, it is not beyond credibility to suggest a connection. Nevertheless, while many Aboriginal men compare the lengths of their subincisions much as Western men might compare penis size, with a larger split indicating increased manliness and sexual potency, and while it does seem that there is at least some sexualized component to having a subincision even if not in actually acquiring one, this assertion strains belief. The resemblance between the penis of a kangaroo and that of a subincised human is slight.

A more plausible theory is that subincision might have originated as a procedure with health benefits. This hypothesis draws inference from similar rituals beyond Australia whose significance is better understood and less shrouded in the mythologies of spiritual or cultural observance. In Brazil, for example, genital surgery has a definitively practical use for the relief of inflammation caused when certain species of parasitic fish enters the urethra. It has been proposed that similar biological relief could well be required were an Aboriginal Australian's urethra to become irritated by foreign bodies. However, the narrowly ritualistic engagement with subincisions across the Australian continent may defy this explanation.

In Fiji and in Tonga, the subincision of the penis is also primarily understood as a quasi-medicinal measure. On these islands, it is perceived that the shedding of blood from the subincision wound functions as a panacea, expelling disease from the body. Similarly, the expulsion of blood as a means of corporeal purification is the underlying motivation behind an analogous procedure undertaken by the Wogeo tribe of New Guinea. Although they supercise rather than subincise the penis, meaning that they open the top rather than the underside of the penile shaft, its purported function is similar to that described by the Fijians. The Wogeo believe that a woman's monthly emission of blood serves to purge her body of pollution, and that it is necessary to incise the penis regularly in order to provide men with a similar outlet. The procedure is also coupled with a heavily symbolic charge, as it is seen to elide sexual differences, simultaneously allowing men to emulate female menstrual bleeding and refashioning the penis to more resemble the vulva. Moreover, subincised men will usually need to squat to urinate, invoking images of sexual similarity.

As Aboriginal Australians have also traditionally construed the nature of menstrual blood as a means for women to purge pollution from their bodies, and given the special attention to the shedding of blood during the arilta rituals, the original use of subincision to reduce sexual difference, symbolically or otherwise, has become a particularly persistent theory on the origin of Australian subincision rituals, especially given the explicitly defined cases elsewhere. In fact, some particular Aboriginal Australian groups use a slang word for vagina when referring to the subincised penis. It is of course possible to produce an overtly Freudian, Oedipal reading of subincision if these symbolic castrations are taken to heart.

The similarities Aboriginal genital rituals share with those from New Guinea have been taken in some academic circles to suggest that the practice was brought to Australia from Guinea in the distant past. Archaeologists believe that the first humans to arrive in Australia did so via a land bridge with New Guinea that existed 40,000 years ago, although it should be noted that subincision in Australia is most concentrated near the center of the continent, and that it is hardly found along the northernmost coast (nearest New Guinea) at all.

In Western culture, certain subsections of the body-modification subculture have appropriated subincision for their own ends. Although not widely practiced, some modifiers who identify with the modern-primitives movement, which emerged in California in the 1980s, have been drawn to the ritualistic aspects of subincision and its tribal associations. Others have been entranced by the possibility of increasing sexual pleasure, as the opening of the urethra produces a new, sensitive genital surface. Subincision was one of the procedures (alongside castration) offered by underground cutters to a predominantly sadomasochistic clientele before the mainstream explosion of body modification in the 1990s, and is now occasionally electively undertaken by interested parties within a broader spectrum of pseudosurgical body-modification technologies.

Further Reading: Bell, Kirsten. "Genital Cutting and Western Discourses on Sexuality." *Medical Anthropology Quarterly (New Series)* 19/2 (June 2005): 125–148; Cawte, J. E. "Further Comment on the Australian Subincision Ceremony." *American Anthropologist (New Series)* 70/ 5 (Oct. 1968): 961–964; Harrington, Charles. "Sexual Differentiation in Socialization and Some Male Genital Mutilations." *American Anthropologist (New Series)* 70/5 (October 1968): 951–956; Larrat, Shannon. *ModCon: The Secret World of Extreme Body Modification.* Toronto: BMEBooks, 2002; Montagu, Ashley. *Coming into Being Amongst the Australian Aboriginies.* 2nd ed. London: Routledge, 1974; Singer, Philip, and Desole, Daniel E. "The Australian Subincision Ceremony Reconsidered: Vaginal Envy or Kangaroo Bifid Penis Envy." *American Anthropologist (New Series)* 69/3/4 (1967): 355–358.

Matthew Lodder

Reproductive System

Cultural History of Menopause

Menopause means a cessation of the menses, or the menstrual cycle. Menopause occurs when the female body no long produces the estrogen necessary for reproduction and the functions of the ovaries cease. Women are born with a limited number of eggs. These eggs are located in the ovaries, which produce the hormones responsible for menstruation. Menopause is a natural part of female reproduction and typically occurs in late middle age. Beyond the inability to produce children, menopause has a number of psychological, social, and cultural implications and meanings, leaving menopause stigmatized by some and honored by others.

Women typically begin menopause between the ages of forty-five and fifty-five. In the Western world, menopause begins on average at age fifty-one. When a woman begins menopause before the age of forty, this can be a result of chemotherapy or a weakened immune system. Smoking, alcohol and drug abuse, obesity, extreme weight loss, physical or psychological trauma, and even nutrition can all influence the onset age of menopause. Recent studies have suggested that, compared to women born early in the twentieth century, modern women are experiencing menopause later in life. Studies also suggest that the most significant factor in determining the age of onset is smoking, with smokers experiencing menopause earlier than non-smokers.

Physically speaking, the female body goes through significant changes with the onset of menopause; the primary characteristics are the decrease of estrogen levels and the cessation of ovulation. Estrogen is a set of female hormones mainly produced by the ovaries, though some estrogens can be produced by other body parts, including adipose tissue, the placenta in pregnancy, and in men, the testes. Estrogen levels begin to increase as young girls reach puberty and begin ovulation, or producing eggs. Importantly, estrogen is responsible for the regulation of premenstrual cycles. Estrogen is also responsible for secondary sex characteristics in women, such as breasts and widened hips. Though estrogen is also prevalent in men, it is released in the male body in very small amounts, to help in the development of sperm.

When menopause occurs, the female body begins to decrease its estrogen production, eventually leaving women infertile. Menopause officially occurs when women have gone through one full year (twelve months) without menstruating. Menopause occurs in three stages, perimenopause, menopause, and postmenopause, although some medical experts also speak of a preceding stage, premenopause. Perimenopause begins when the ovaries begin to produce less estrogen; in the West, the average age of onset is forty-seven. Perimenopause can begin up to ten years before menopause begins, ending when the ovaries no longer produce eggs. Menopause occurs when the body ceases to menstruate and is associated with a range of symptoms. Postmenopause describes the period after menopause ends.

Most menopausal women go through a number of physiological changes beyond their inability to have children. The decreases in hormones have a major impact on the body; however, the symptoms vary by ethnicity and nationality. In the United States, for example, women complain more of hot flashes, while in Asia hot flashes are reported less often. Asian women, however, report joint pain more frequently. Symptoms that are often associated with the onset of menopause include: physiological symptoms such as irregular periods (shorter, longer, lighter, or heavier), hot flashes (short bursts of intense heat that last anywhere from one to four minutes that can cause higher skin and body temperature as well as an increased heart rate), night sweats or other periods of vasomotor instability, sudden tear production, dryness of the vagina from a thinned vaginal lining, insomnia, fatigue, itching, headaches and migraines, nausea, bloating, weight gain, weakness of the bones (osteoporosis) and fingernails, frequent urination, soreness of muscles and joints, increased facial hair, and hair loss. Some of the psychological symptoms include mood swings and irritability, anxiety, depression, memory lapses, decreased libido, and dizziness.

Menopause can also greatly impact a woman's sex life. With a decrease in hormones, which often results in a decreased libido, coupled with other more physical symptoms such as vaginal dryness, women may lose their sexual desire and may find sex to be quite painful. These symptoms could last several years as the body readjusts to its new hormone levels. Importantly, the onset of menopause coincides with other issues women tend to go through later in life, such as the death of a spouse or parent. As a result, some of the symptoms may not necessarily come from the onset of menopause itself, but the hormonal changes coupled with lifestyle changes, which may increase the difficulty that some women may experience with menopause.

Menopause and Medicine

The word "menopause" (initially, *ménèspausie*) was coined by French physician C. P. L. de Gardanne in the early nineteenth century, first appearing in 1812 in his dissertation on the subject. It was not until the eighteenth century that menopause became a popular subject in medical literature. Before then, menopause was not often publicly discussed, and when it was, it was often simply referred to as "the change." But in early modern Europe, menopause became highly medicalized.

Prior to the eighteenth century, there was little discussion of the ceasing of menses as a medical problem. Classical medical literature rarely discussed menopause. However, menstrual periods were long seen as cleansing the body. Hippocrates, the Greek physician from the island of Cos, born around 460 BCE and known as the "father of medicine," believed that women's bodies were cleansed during their menstrual cycles, when all of the impurities went into the bloodstream and were released. Menstrual blood was often viewed as tainted and polluted, and menstruation was seen as necessary to rid the body of impurities and to keep the female body functioning properly. Early modern physicians in the seventeenth through nineteenth centuries also believed that menstruating was vital to health, and that irregular periods could cause significant illnesses. Seventeenth- and eighteenth-century texts describe menopause as the result of perceived processes of aging, such as the body's weakening ability to excrete menstrual blood, or the thickening of the blood. Because her periods ceased, the menopausal woman was thought to be accumulating surplus blood, or experiencing "humoral corruption," linked to abdominal pain, fatness, and excess heat. An alternative theory of "irritation" framed menopause not as a failure of the evacuation of blood from the body but as a period of intensification of complaints common during the menstruating years.

Physicians of the eighteenth and nineteenth centuries treated menopause as highly problematic. In 1710, the first known monograph on menopause, written by Simon David Titius, was published. The volume focused on the negative effects of the process on women's health, and marked the beginning of an intense period of medical interest on the subject. Popular manuals on how to manage menstruation were first released in the late eighteenth century, and a variety of preparations to treat symptoms became widely popular. It has been noted that early every conceivable illness was linked to menopause. Menopause was perceived by many physicians as well as women themselves as dangerous for women's health, including their mental health. Many medical practitioners believed that sexual desire decreased, that menopause caused the vagina and breasts to shrink and shrivel, and that menopausal women became more masculine. A menopausal woman would gain weight, go bald while simultaneously growing facial hair, her voice would deepen, and her clitoris would enlarge. Furthermore, the postmenopausal woman was morally and mentally suspect. She could be considered overly licentious, polluted, or even susceptible to envy and bitterness at the loss of her fertility. Alternatively, while in the "irritation" menopause itself was highly pathologized, the postmenopausal period was seen as one of increasing order and stability in a woman's life.

While many physicians of the eighteenth and nineteenth centuries saw menopause as a medical problem, other physicians argued that menopause was a perfectly natural event that should cause no upset. They pointed to rural women who were, they argued, perfectly willing to accept "the change" without complaint. These writers saw the real cause of problems associated with menopause to be the overstimulating lifestyles of the urban middle class.

Thus, during this period, menopause was both medicalized, or treated as a medical problem, and demedicalized, considered a "natural" process that should not demand medical attention.

This paradox persisted in the twentieth century, when menopause was alternatively viewed in medical literature as a normal process and as a pathological state. The medical approach to menopause in the twentieth century was initially conservative because doctors believed it was a natural process that did not need intervention. Early hormone treatments, known as "organotherapy," emerged between 1897 and 1937; however, it was not until the development of the drug DES (diethylstilbestrol) in 1938 that hormone therapy was legitimized. Up until the mid-twentieth century, however, doctors were reluctant to treat symptoms of menopause. In the 1960s, Robert A. Wilson, a gynecologist, and his wife Thelma Wilson argued that menopause should be treated as an estrogen-deficiency disease that is highly damaging to women's lives, creates significant psychological distress, and must be treated medically. The publication of Robert Wilson's book *Feminine Forever* (1966) sparked widespread discussion, but no consensus, on whether menopause should be treated medically.

The feminist movement in the 1960s also brought menopause and other women's health issues into public focus. While some feminists were critical of the medicalization of menopause, which treats the process as pathological and in need of intervention, some would argue that the greater willingness of women to demand medical attention for menopause is partly responsible for a shift in the medical attitude toward menopause. By the start of the 1990s, doctors were widely prescribing hormone replacement therapy (HRT). Initially, these replacements consisted only of estrogen; however, progesterone was eventually added. HRT is used to decrease the symptoms of menopause such as hot flashes, depression, and vaginal dryness. Medical professionals also suggest that HRT strengthens hair, bones, and even increases brain functioning.

In the late 1990s, HRT was one of the most widely prescribed drugs in the United States. However, the drug has been found to have negative side effects and significant risks. In the 1970s, during early studies on hormone replacements, researchers soon found that these replacements increased rates of uterine cancer. However, in 2002, researchers found that women who were on hormone replacement therapy had higher risks for cancer, strokes, and heart disease. Since then, HRT use has become highly controversial.

Menopause and Culture

Women have a variety of different reactions to the physical and psychological changes of menopause; culture and lifestyle often influence how a woman will experience menopausal symptoms. For example, women sometimes enter into therapy and they may take hormone replacements to treat symptoms of menopause, while others treat menopause as a nonmedical event or as a celebrated time in one's life. Depending on one's culture, menopause also takes on varied symbolic significance.

A majority of the research on menopause tends to report on the negative aspects of menopause, such as infertility, decreased sex drive, and hot flashes.

However, women have reported some positive aspects of menopause. Some welcome the cessation of fertility because it eases the problems of birth control. Others are glad to see the end to the pain and inconvenience of menstrual cycles. Research also suggests that women's experiences with menopause vary based on culture. Growing old, aging, and notions of beauty differ based on culture, and these can significantly influence the psychological aspects of menopause. Furthermore, physical symptoms of menopause vary by nationality, ethnicity, and even class.

Some reports suggest that, in societies where aging is frowned upon and is met with a decrease in power and privileges, women tend to view menopause more negatively; conversely, in societies where there is less stigma to aging, women experience menopause more positively. Women in Asian countries tend to fare better with menopause, reporting fewer negative symptoms than women in the West. In some Asian cultures, women gain increased power and privileges as they age, while aging is associated with loss of status in the United States. In some Native American tribes such as the Lakota, Plains Cree, and Navajo tribes, menopause is viewed in a positive light, with women gaining increased respect and power after its onset. Some of the cultures in which women gain power with the onset of menopause believe that women lose their femininity with menopause and, as a result, are perceived as becoming more intelligent.

Interestingly, a number of the physical symptoms menopausal women experience can vary based on culture and even location. For example, one 2005 study compared the experiences of menopause in a sample of women in Papua New Guinea to a sample of women in Germany. The researchers found that the different groups of women experienced different physical symptoms, with women in Germany expecting a higher likelihood of hot flashes and women in Papua New Guinea expecting a greater possibility of sexual problems and vaginal dryness. The researchers found that the experiences of menopausal women in these different regions of the world followed suit with their expectations. In another study, women in Nigeria reported little bloating but greater incidences of headaches. Women in Japan reported very few symptoms such as hot flashes and bloating. In this same study, however, women in the United States reported very high rates of bloating, but very few headaches.

Experiences with menopause vary not only by culture, but also by sexual orientation. A recent study examining the experiences of menopause pointed out that most literature on menopause tends to focus on heterosexual women. However, this study compared the menopause experiences of lesbians and heterosexual women and found that women with female partners tend to be more open than heterosexual women about their sexual needs, which had a positive impact on their menopausal and postmenopausal sex lives.

Social class, or one's social economic group, even has an influence on how women experience menopausal symptoms. For example, women from a higher socioeconomic class as well as women who work outside the home tend to report fewer hot flashes than those from lower economic groups and those who work inside the home.

Menopause in Popular Culture

While menopause was once a taboo subject, it is now fodder for public discussion. As women live longer, menopause, its symptoms, and issues, have been gaining increased publicity in popular culture and the media. Importantly, open discussions of menopause are not as taboo as they once were. Openly addressing menopause not only confronts issues of sexuality, but also the issues associated with growing older.

Whereas menopause was once known as simply "the change," women in Western nations are becoming more comfortable with talking about it. Some are becoming empowered with the onset of menopause. Women have written poems, jokes, and have even produced art meant to empower those going through menopause. There have been many instances and representations of menopause on television, talk shows, and movies. For example, in recent years in the U.S. media, talk shows such as *The View* and well as newsmagazine shows have featured many stories on issues pertaining to menopause, including the controversies surrounding hormone replacements. Sitcoms ranging from the British show *Absolutely Fabulous* to 1980s classics such as *The Golden Girls* and *The Cosby Show* all featured characters going through menopause, and the reactions of their friends and loved ones.

People have even been able to view menopause lightly, with Irish singer and comedian Cahal Dunne writing and performing the song "Here Comes Menopause" to the tune of "Here Comes Santa Clause." The well-received comedy musical, *Menopause, the Musical*, written and produced by Jeanie Linders, has been touring the United States, Great Britain, Canada, and Australia since 2001.

Further Reading: Delaney, Janice, Mary Jane Lupton, and Emily Toth. *The Curse: A Cultural History of Menstruation*. New York: New American Library, 1977; Griffen, Joyce. "A Cross-Cultural Investigation of Behavioral Changes at Menopause." *Social Science Journal* 14 (1977): 49–55; Houck, Judith A. *Hot and Bothered: Women, Medicine, and Menopause in Modern America*. Cambridge, MA: Harvard University Press, 2006; Janiger, O., R. Riffenburgh, and R. Kersh. "Cross-cultural study of premenstrual symptoms." *Psychosomatics* 13 (1972): 226–235; Kowalcek, I., D. Rotte, C. Banz, K. Diedrich. "Cross-cultural and intra-cultural comparison of pre-menopausal and post-menopausal women in Germany and in Papua New Guinea." *Maturitas*, vol. 51, no. 3 (2005): 227–235; Martin, Emily. *The Woman in the Body: A Cultural Analysis of Reproduction*. Boston, MA: Beacon Press, 1987; Stolberg, Michael. "A Woman's Hell? Medical Perceptions of Menopause in Preindustrial Europe." *Bulletin of the History of Medicine* 73.3 (1999): 404–428; Winterich, Julie A. "Sex, Menopause, and Culture: Sexual Orientation and the Meaning of Menopause for Women's Sex Lives." *Gender and Society* 17, 4 (2003): 627–642; Voda, Ann M. *Changing Perspectives on Menopause*. Austin: University of Texas Press, 1982.

Angelique C. Harris

Cultural History of Menstruation

Menstruation refers to one aspect of the female reproductive cycle, which commences at puberty and ceases at menopause. It is the most visible element of the fertility cycle, as the blood which temporarily lined the uterus—in preparation to nourish a fertilized egg—is shed, seeps through the cervix, flows down the vagina, and emerges from the genital area. This flow of blood usually indicates that a woman is fertile, though not pregnant, but that she may become so as her cycle commences again. The first day of bleeding is generally counted as day one of the menstrual/fertility cycle. However, as with the human body and its processes more generally, along with sex, gender, and sexuality in particular, this physiological process is subject to complex belief systems and value judgments that are evident in spiritual, social, and medical practices across the globe and throughout history. Indeed, that this periodic bleeding is specific to females, and is indicative of their reproductive potential, has made it one of the most highly symbolic of all bodily functions, affording it a unique status—occasionally positive, but more commonly abhorrent—which bears no comparison with responses to other bodily fluids, such as sweat, tears, saliva, semen, or nonmenstrual blood.

Menstruation seems to occupy the most positive status in ancient matrifocal, or woman-centered, cultures where it is likely that the relationship between heterosexual intercourse and reproduction was not clearly established. Here, women's capacity to give birth, as well as the mysterious bleeding which visited them in cycles and which corresponded with, or could be measured by, the phases of the moon, inspired awe and reverence, generating many mystical and metaphorical associations between women's blood, reproduction, life, death, and the divine. The period of blood shedding was often believed to indicate a woman was at the height of her powers, possessing supernatural, curative, or creative abilities.

Possible remnants of belief systems which reified menstruation (and the female body) were evident up until fairly recently in cultures where the high status of menstruation, and the qualities which women were seen to embody as a result of their periodic bleeding, engendered rituals in which men emulated this blood shedding. In Papua New Guinea, men routinely induced nasal bleeding in order to endow them with the health, strength, and attractiveness, as well as the ritual purification, which menstruation was seen to afford women. And, this "Island of Menstruating Men," as it became known, is not unique. In other areas of New Guinea, as well as among the Australian Aboriginal people, male bleeding is also produced, but this time by cutting directly into the penis. Here, not only is the flow of blood seen to emulate menstruation, but the incisions made, which cut through to the urethra, create an opening in the penis of a similar shape to the female genitals. Further, this practice is such that males are no longer able to urinate through the tip of their penis, and are thus required to adopt a female-like squatting position. In addition, among some Australian tribes, where red ochre has much ritual and symbolic importance, the substance is believed to originate from the menstrual blood shed by mythical women.

In some cultures, the onset of a girl's first menstruation is a time of celebration to mark her entry to womanhood. The celebrations can be simple affairs primarily for close kin (South India), or elaborate rituals that last for many days and include the whole community (the Ashanti, Wogeo, and Nowago peoples). In many traditional African cultures, where scarification is a sacred practice, a girl's first period defines the point when she will take on her first set of ritually inscribed scars and enter into her new, adult status.

The majority of cultural evidence, however, points to much less-favorable attitudes toward menstrual bleeding. This is evident in the more negative spiritual beliefs and codes of practice around menstruation that are broadly termed as "menstrual taboos," as well as within cotemporary cultural and social mores regarding the practicalities of dealing with menstrual blood. Negative attitudes to menstruation are also endemic to medical norms that pathologize not only the blood itself, but also the entire female reproductive system, and, as a result, the female gender per se.

Menstrual taboos have been widely documented by anthropologists studying small-scale cultures, and these taboos are integral to the broader codes of gender behavior. The taboos generally center on beliefs that women are in some way dangerous, polluting, or threatening to men and masculinity when they are bleeding. Common themes refer to menstruating women, simply by coming into contact with them, blunting tools and weapons, wilting crops, turning food sour, or weakening men's potency. Thus, menstruating women have been required to observe stringent protocols in order to prevent contamination or damage. These practices have ranged from the symbolic, such as wearing a red headband to warn others of her potential danger (Angola), to the physically demanding and brutal. In many cultures, women have been sent into exclusion in special huts away from their community for the duration of their bleeding. While in some regions these may be seen to create special women's spaces and to afford some kind of female solidarity, in other societies, such as the Inuit, women have endured enforced isolation and food deprivation while occupying these menstrual huts. Other cultures (such as the Tincunas) required flagellation for menstruating women alongside the plucking out of her hair; while in other societies, a whipping was applied if a woman's period lasted for more than the prescribed amount of days (Persia). In some cultures, punishment by death was prescribed for a woman who failed to properly adhere to the menstrual codes of practice and thereby risked the well-being of her entire community, as was the case with the native tribes of the Illinois region. In other cultures, the practices that menstruating women have endured to avert contamination sometimes inadvertently resulted in death. This was common among the Guarians, where menstruating women were sewn into hammocks and suspended over a fire—often for days at-a-time—to ensure their purification.

Menstrual taboos are not the sole preserve of small-scale cultures, and are prevalent among many belief systems that continue to exist in large-scale contemporary cultures. Twentieth-century Greek Orthodox Christians prohibited women who were menstruating from receiving Holy Communion because of their impurity. Menstrual taboos have also been found to have a more sinister presence in earlier Christian history. Witch trials, which took place across Europe throughout the Middle Ages, and during which around eight million women were tortured and killed, could be indicative of a profound menstrual taboo. This persecution of primarily unmarried women could be indicative of the Christian Church's fear of unrestrained female sexuality but, more specifically, that the qualities ascribed to witches were identical to the taboos which were thought to be signified by menstruation. These included unbridled female sexuality; the specific, impure, and possibly dangerous qualities of the blood; emotional and physical unpredictability; inexplicable powers over crops and people (witches were often herbalists and/or midwifes); and posing a threat to men, masculine power, and purity.

Other contemporary religions also enact—though much less drastically—menstrual stigma and taboo. In some strands of Judaism, women observe the prohibition of *niddah* during menstruation. This prohibition dates back to the early scriptures and was primarily intended to prevent women entering the temple while bleeding. However, observing niddah includes other prohibitions (with some variation across regions and time periods) that generally involve restrictions around the handling of food, entertaining certain company, and abstaining from sexual intercourse. Another Jewish tradition, still widely practiced today, dictates that when a girl announces the onset of her first period to her mother, she receive a sharp slap across her face. Menstruation has a further connection with Judaism, in relation to the persecution of Jews in Europe during the seventeenth century. Jewish people were identified as particularly abhorrent or inhuman on account of the widespread belief that Jewish men as well as women menstruated.

Menstrual taboos perpetuate many cultural and spiritual belief systems, and generally stigmatize both the periodic bleeding as well as the female body. However, scientific belief systems are not immune to such prejudices, and similar views of menstruation can be found within historical and contemporary Western medical practice.

The female reproductive system and its functions, particularly menstrual bleeding and the symptoms associated with it, have been considered evidence of female instability and inferiority throughout medical history. However, fluctuations in medical interest occur and, over time, the focus has shifted between concerns regarding the *physical* aspects of blood shedding and its implications for female fragility and vulnerability, as well disorders such as amenorrhea (the absence of menstruation) and dysmenorrhea (heavy, painful, or irregular bleeding), and the *emotional* and *psychological* imbalances caused by the menstrual cycle, such as premenstrual syndrome. Such medical evidence legitimated the exclusion of women from education in the eighteenth century, from practicing medicine in the nineteenth century, and from joining military service in the early twentieth century. However, these menstrual problems and prohibitions are apparent only within women of certain social classes and ethic groups (white, middle-, and upper-class women), since working-class, black, and ethnic minority women have never been considered so debilitated by menstruation as to prevent them from working. Nor does the "biology" of menstruation prevent women from participating in exclusively male activities during national emergencies such as wartime.

The medical view of menstruation as a potentially debilitating problem is also reflected in the marketing of contemporary menstrual products. Sanitary towels and tampons are marketed as forms of "protection," as a hygienic or liberating response to a bodily function that is largely represented as a burden, and indeed, which is commonly dubbed "the curse." Here, menstruation is defined not only as a physiological problem, but also as a social aberration with similar qualities to a taboo. Thus, menstrual products focus on possible negative aspects of menstruation including, leakage, odor, and unattractiveness—particularly to men. It is depicted as antithetical to femininity and, as such, as something that requires rigorous, specialist, and expensive interventions. Menstrual products largely define menstruation as shameful, and focus on the threat and stigma of one's menstrual status being exposed.

This pathologization of menstruation in culture, religion, and medicine has led many feminists to cultivate a countercultural, body-positive attitude toward periods, reviving ancient matrifocal beliefs and practices. This reversal of the stigmatizing evaluations of menstruation as taboo includes a more open acknowledgment and celebratory attitude to this "special time," especially at a girl's first period. Feminists have also encouraged the boycott of commercial menstrual products, which depend upon and further both the medicalization of women's bodies and the shame and stigma of menstruation. They support the use of environmentally sound and woman-friendly, reusable products such as the Keeper, the Moon Cup, and Sea Sponges, as well as homemade menstrual products. Some feminists have also advocated the artistic, political, and creative uses of menstrual blood in both the public and private realms. Indeed, the cultural response to these explorations, including in the works of the artists Kathy Acker and Judy Chicago, is as clearly a demonstration of the status of menstruation in contemporary culture as the artwork itself. *See also* Penis: Subincision.

Further Reading: Bettleheim, Bruno. *Symbolic Wounds: Puberty Rites and the Envious Male.* London: Thames & Hudson, 1955; Favazza, Armando, R. *Bodies under Siege: Self-mutilation and Body Modification in Culture and Psychiatry.* 2nd ed. Baltimore, MD: The John Hopkins University Press, 1996; Houppert, Karen. *The Curse: Confronting the Last Taboo: Menstruation.* London: Profile Books, 2000; Laws, Sophie. *Issues of Blood: The Politics of Menstruation.* London: Macmillan Press, 1990; Muscio, Inga. *Cunt: A Declaration of Independence.* 2nd ed. Emeryville, CA: Seal Press, 2002; Shail, Andrew, and Gillian Howie, eds. *Menstruation: A Cultural History.* New York: Palgrave, 2005; Shuttle, Penelope, and Peter Redgrove. *The Wise Wound: Menstruation and Everywoman.* 2nd ed. London: Marion Boyars, 1999; Weideger, Paula. *Female Cycles.* London: The Women's Press, 1978.

Kay Inckle

Fertility Treatments

Fertility treatments are practices that aim to facilitate conception and pregnancy. Attempts to enhance (or prevent) fertility have a long history, and across cultures, a wide range of foods, herbs, objects, and practices have been advocated and enacted either to prevent conception, terminate a pregnancy, or treat the failure to conceive or carry a pregnancy to term. Variations in dominant interventions over time and across different cultural contexts are determined largely by the prevailing theories of the reproductive body. For example, in the West, bloodletting, herbs, and tonics were advocated well into the nineteenth century in order to correct bodily imbalance, while exercise and electric shocks were used to stimulate menstruation in those deemed to be retaining menstrual blood. Infertility has also been attributed at various times and places to diabolic intervention, or to divine retribution, inviting religious interventions or rituals of cleansing. Significantly, the vast majority of

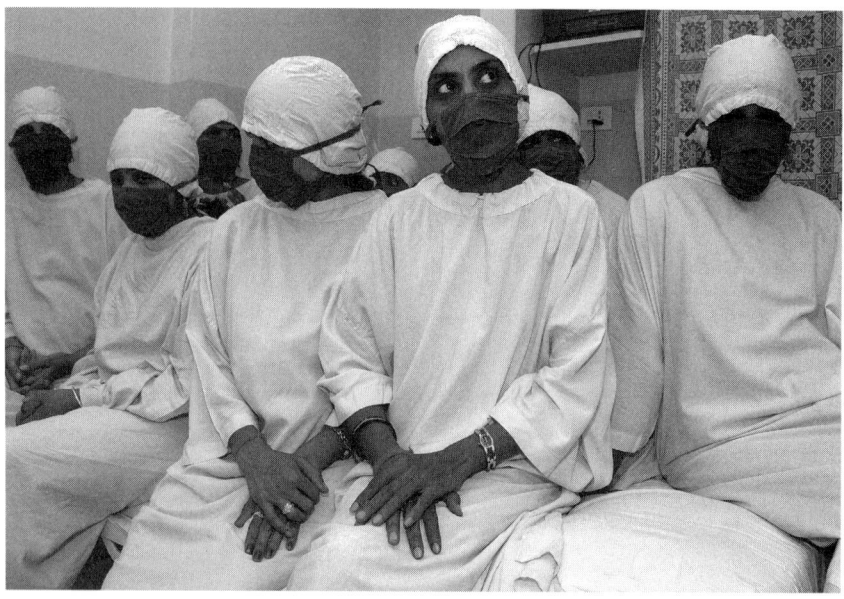

Surrogate mothers look on during an interaction with the media at the Kaival Hospital at Anand, India, 2006. Surrogate mothers there get between Rs. 100,000 (US$2,222) to Rs. 200,000 (US$4,444) for surrogacy. (AP Photo/Ajit Solanki)

fertility treatments across time and place (including the present) share a focus on the female body as the site of infertility.

Premodern Fertility Treatments

Historically and across cultures, a wealth of premodern rites, rituals, and practices aimed to enhance fertility. There are many gods and goddesses associated with fertility in polytheistic cultures, including Cybele, the Phrygian goddess who was worshipped in various forms from Neolithic times through the Roman Empire. There are also the Hellenistic figures Aphrodite, Hera, and Adonis, the Roman goddesses Venus and Juno, Hathor of ancient Egypt, the Inuit goddess Pukkeenegak, the Aztec goddess Tlazolteotl, the Hindu goddess Parvati, and the Celtic figure Brigid, among others. The ancient Western European traditions of maypole dancing and English Morris dancing are both believed to have their origins in fertility rites, both for the people participating in them, and for the land on which they lived and farmed. Early modern Europeans also used a variety of plants, potions, and amulets to boost fertility (for example, coriander to boost male fertility, and the wearing of lodestone amulets for women). China has a long history of the use of herbs, acupuncture, and other traditional techniques for treating infertility. In Ghana and other parts of Africa, carved fertility dolls worn around the neck, or positioned next to the marital bed, are traditionally believed to ensure the birth of a healthy baby. Asante Akua'ba dolls are well-known symbols of fertility; they are usually carved from wood and have disk-shaped heads, and they often have extended arms and a base instead of legs. In Bengal (Bangladesh and the state of West Bengal in India), the plantain plant has long been associated with fertility. In India, trees have been symbolically important in wedding rituals for their association with fertility rites.

Contemporary Western practices in nonmedical settings borrow from indigenous medicines and methods and include acupuncture, diet, herbal treatments, meditation, hypnosis, yoga, and massage to boost fertility. Many people wear fertility amulets, and Westerners have increasingly been interested in reviving other traditional practices.

Contemporary Biomedicine

Contemporary biomedicine has supplanted many of these bodily paradigms and their associated interventions, although even in those countries where biomedicine is the dominant voice in relation to health and illness, many of these practices remain current within alternative, complementary, or traditional medicine. In China, for example, the use of traditional herbal medicine remains popular in the treatment of infertility.

The late twentieth and early twenty-first centuries have seen the dramatic proliferation of biomedical fertility treatments, or assisted reproductive technologies (ARTs). These include: insemination techniques where sperm is artificially introduced into the vagina (as with donor insemination [DI]), or transferred directly to the uterus as in intrauterine insemination [IUI]); the use of hormonal drugs to stimulate ovulation; and in vitro fertilization (IVF) techniques, in which eggs are fertilized outside of the body and subsequent embryos are selected for transfer to the woman's body. These procedures, and especially those reliant on IVF, are rarely out of the media spotlight, and have considerable social and cultural significance. Therefore, these fertility treatments will be the focus of this entry. The rise of IVF represents the contemporary epitome of fertility treatments, and has provoked considerable opposition and concern. These treatments are a global phenomenon, both in terms of the different ways in which those technologies have been accepted, and also in the global circulation of reproductive resources.

Test-Tube Babies

The world's first IVF baby, Louise Brown, was born in 1978 to massive media fanfare. Her birth marked the culmination of years of scientific efforts to understand, simulate, and manage the "natural" reproductive process. Although this was not the first IVF pregnancy (it occurred in 1977, but was ectopic and had to be terminated), the successful live birth became a key landmark in the history of fertility treatments. The birth made stars of IVF pioneers Patrick Steptoe and Robert Edwards, and it was the picture of them (rather than the parents) holding the newborn Louise that became one the iconic images of IVF. Despite her own lack of enthusiasm for the media spotlight, Louise herself has remained an obsessive focus of media attention, and the birth of her son in December 2006 was reported globally.

Louise's birth in 1978 offered hope to those struggling to conceive, and in particular, to women with blocked fallopian tubes, since IVF removed the need for an egg to travel along the fallopian tubes toward the uterus. Since then, technological developments have massively expanded the potential pool of IVF patients. While the first IVF cycles involved a single egg harvested during the woman's natural cycle, techniques of drug-induced super-ovulation followed in the early 1980s, enabling doctors to collect and fertilize multiple eggs in a single treatment cycle. This was followed in the early 1990s with a technique for injecting a single sperm into an egg (intracytoplasmic sperm injection, or ICSI), further expanding the application of IVF to those affected by male factor infertility. Accompanying techniques of the cryopreservation of IVF embryos enabled those undergoing treatment to have multiple embryo transfers without having to go through the whole process of ovarian stimulation, egg collection, and fertilization each time. These developments have led to the creation of a thriving biomedical fertility industry, and in June 2006, it was announced at the European Society of Human Reproduction and Embryology (ESHRE) annual conference in Prague that since Louise Brown's birth, an estimated 3 million babies have been born using IVF and its associated technologies, and an estimated 1 million IVF cycles are performed globally each year. However, the failure rates for IVF remain high, and the "miracle baby" of IVF will remain elusive for the majority of those undergoing treatment. Fertility treatment is also increasingly linked with genetics research and imagined future stem cell therapies. Human eggs and embryos are in high demand in the field of stem cell research, and these can only be acquired using the techniques and practices of fertility treatment. The fact that each IVF cycle frequently generates embryos in excess of those required for treatment means that patients who choose not to freeze embryos for later use are increasingly being asked to donate those embryos to stem cell research programs in some countries. Furthermore, in the United Kingdom, the regulatory body for fertility treatment, the Human Fertilisation and Embryology Authority (HFEA), recently granted licenses for "egg sharing" agreements between women and research scientists. These would enable women to receive subsidized IVF treatment in exchange for a portion of the eggs collected.

Oppositional Voices

While IVF and its associated fertility treatments are a potent source of hope for those experiencing infertility, the birth of the first "test-tube baby" and the subsequent expansion of fertility treatments also triggered an ongoing flood of vehement and passionately oppositional voices. This opposition focuses on four key sets of issues: "pro-life" debates; disruption of the "natural" reproductive and social order; "designer babies" and what it means to be human; and feminist responses in relation to the health and rights of women.

"Pro-life" Debates

As has already been discussed, IVF and its associated technologies frequently produce embryos in excess of those needed for treatment. Unused embryos either can be frozen for future use, donated for research or for another's fertility treatment, or discarded. For those holding a strict "pro-life" position that treats the IVF embryo as having a "life" status equivalent to that of a baby, IVF is an immoral practice that routinely creates and then destroys human beings. Consequently, fertility treatments such as IVF have been drawn into wider debates around abortion and the "right to life." A striking example of this is the so-called "snowflake children" in the United States. These babies originate from frozen IVF embryos that are no longer needed for treatment and are subsequently "adopted" and used by another couple. U.S. President George W. Bush is regularly photographed with "snowflake children" as a means of affirming not only his opposition to abortion (and the discarding of excess IVF embryos), but also to the use of embryos for stem cell research.

Disrupting the Reproductive Order

One of the consequences of the proliferation of fertility treatments in the late twentieth and early twenty-first centuries is the opening up of the possibility of new and difficult-to-categorize family forms and kinship relations. The contracting of surrogates to carry a pregnancy and the use of donor eggs and sperm fragment the previously unitary categories of mother and father into genetic, gestational, and social roles. A child created from a donor egg, gestated by a surrogate, and raised by a third woman could be described as having three mothers; the legal wrangles over who constitutes the "real" mother in surrogacy cases where the gestational mother has refused to relinquish the baby highlight the complexity of these categorical fragmentations. Similarly, cases of egg donation and surrogacy between family members across generational boundaries (for example, if a mother gestates a pregnancy for her daughter with IVF embryos created using the daughter's eggs) disrupt the kinship categories of grandmother, mother, daughter, and sister in new and challenging ways. For many people, these (albeit relatively unusual) cases signal a troubling disruption of the "natural" reproductive order and of the institution of the family.

The "family" in these oppositional accounts is normatively defined in heterosexual terms, and another common source of disquiet about fertility treatments is the extent to which they have created new opportunities for people who are single or in same-sex relationships to become biological and/or social parents. Donor insemination (DI) has a much longer history than treatments such as IVF and can easily be practiced through privately negotiated arrangements between gamete donors, recipients, and surrogates, facilitating reproduction outside of the surveillance of the biomedical domain. This has also led to the creation of new parenting relationships, with multiple parents (biological, gestational, and social) sometimes negotiating in advance their degree of future involvement in the life of the child. IVF, however, remains inescapably within the biomedical domain and is therefore less subject to individual control; nevertheless, in many countries, for those who can afford it and where the law permits it, IVF and its associated technologies offer both male and female same-sex couples new opportunities to become parents. While some have celebrated this expansion of reproductive opportunities, those who hold in place the primacy of the heterosexual nuclear family as the normative reproductive standard see it as deeply problematic.

Meddling with Nature

Many of the concerns relating to the new familial and kinship arrangements that have arisen from fertility treatments are grounded in the perceived tension between reproduction as a "natural" process and technological interventions into those processes. While IVF is frequently represented as giving nature a "helping hand," concerns that it reflects attempts to "meddle with nature" are expressed through three key tropes which have come to connote a bioscientific challenge to what it means to be human.

The first of these is the epithet, Frankenstein—as in the tabloid media phrase, "Frankenstein technologies." The name is taken from Mary Shelley's hubristic scientist, Victor Frankenstein, and has come to connote monstrosity and scientific misadventure (although in the novel, it is social exclusion that drives Frankenstein's "creature" to violence, rather than being an inherent social threat.) The second key trope through which concern about fertility treatments are expressed is through the Aldous Huxley's 1932 novel, *Brave New World*, with its potent image of babies in bottles, genetically engineered with specific traits in order to create and sustain a deliberately stratified and pacified society. The phrase "brave new world" is mobilized ironically in media debates around reproductive technologies to signify a dystopian vision of lost individuality, reflecting fears around cloning and the ability to intervene in the body at the genetic level, potentially transforming what it means to be human. This relates closely to the third, and perhaps dominant trope, through which these concerns are expressed: the "designer baby." While at present it is not possible to "design" babies at the genetic level (or, indeed, grow them in bottles), technologies of genetic selection such as preimplantation genetic diagnosis (PGD), in which embryos created through IVF are genetically tested (usually for specific diseases) prior to embryo transfer, have been easy targets of the "designer baby" epithet. This is particularly likely when the grounds for selection are controversial. One example of this is the use of PGD to select for sex. This is often done for medical reasons; for example, many X-linked diseases cannot be tested for specifically, but predominantly affect boys, so those who know they are at risk may choose to exclude all male embryos. But there is also a considerable demand for social sex selection, either in response to social preferences for one sex (usually boys) over the other, or to the desire to have children of both sexes within the family (what is often referred to as "family balancing"). Another example is the phenomenon of "savior siblings," in which PGD is used to select embryos that are a tissue match for a living sick child, making the subsequent baby a perfect donor of stem cells from cord blood and bone marrow. The "designer baby" label easily attaches to these cases, reflecting fears about both "meddling with nature" and the potential commodification of human life that these new choices can be seen to represent.

Feminist Responses

The presumed centrality of reproduction and motherhood to normative femininity, as well as the fact that fertility treatments focus almost exclusively on women's bodies, mean that fertility treatment is a long-standing source of concern and interest among feminists. These concerns include: the assumptions built into the treatments and their delivery that women *should* have children, and should be prepared to go to great lengths to do so; fears about the risks that fertility treatments pose to women, for example, in relation to the use of fertility drugs whose long-term effects are not yet known; concerns about the exclusivity of fertility treatments, access to which is often constrained by financial limitations and the extent to which the desired family conforms to social and cultural norms; and the ways in which the focus on fertility treatments detracts from the prevention and treatment of fertility-damaging disease. Indeed, many feminists have highlighted the extent to which fertility treatment does not *treat* the causes of infertility at all, but instead circumvents them.

In the 1980s, as the rise of IVF began to gain pace, a radical feminist group called the Feminist International Network of Resistance to Reproductive and Genetic Engineering (FINRRAGE) emerged as a strongly oppositional voice to fertility treatments. Its members argued that reproductive technologies in general (including contraception and birth technologies), and fertility treatments in particular, constituted the experimentation on, and exploitation of, women's bodies. FINRRAGE claimed that those technologies are irretrievably patriarchal and damaging to women, both individually and as a class. They also highlighted the racialized and classed underpinnings of fertility treatments at the national and international levels. However, FINRRAGE was never able to take into account women's desire to engage with those technologies outside of a framework of either duped passivity or traitorous complicity. More recently, feminists have focused on the complex ways in which women engage with fertility treatments, negotiating risk, responsibility, and desire within the social structures of gender, age, race, class, and so on.

Global Issues

At the ESHRE conference in Prague in 2006, it was reported that almost half of all IVF cycles took place in only four countries: the United States, Germany, France, and the United Kingdom. However, while financial constraints and religious and legal restrictions impact the availability of fertility treatments in different countries, IVF and its associated fertility treatments are a global phenomenon, although their social and cultural significance and use varies enormously across different contexts. In China, for example, in the context of a single-child policy, IVF is conceptualized as a means of achieving a "quality" child, and in strongly pronatalist cultures, fertility treatments offer women the hope not only of a much-desired baby, but also of avoiding the social exclusion of childlessness. In many poorer countries, fertility services may simply not exist, with international donors and national governments focusing instead on population control, even though poverty and poor primary health care is strongly associated with fertility-damaging disease.

Looking at the global context, it becomes clear that, although on one level, fertility treatment is a relatively private matter between a person or couple and their doctors, it also operates within a worldwide nexus of complex relations of gender, race, and class, with services and biological materials moving fluidly across national boundaries. So-called "fertility tourists" from developed countries travel overseas in search of cheaper treatments, and those living in countries where fertility procedures such as IVF are illegal (or not widely available) travel overseas for treatment (if they can afford it and are able to). There is also an extensive international trade in eggs for fertility treatment, with poorer women selling eggs through intermediary clinics for use by wealthier women in other countries. The stem cell research industry has further expanded this international dimension, and stem cell lines grown from donated excess IVF embryos are banked and sold globally for research.

Further Reading: Challoner, Jack. *The Baby Makers: The History of Artificial Conception.* London: Channel 4 Books, 1999; Handwerker, Lisa. "Health Commodification and the Body Politic: The Example of Female Infertility in Modern China." In Anne Donchin and Laura Purdy, eds. *Embodying Bioethics: Recent Feminist Advances.* Lanham, MD: Rowman & Littlefield, 1999; Hogben, Susan, and Justine Coupland. "Egg Seeks Sperm: End of Story? Articulating Gay Parenting in Small Ads for Reproductive Partners." *Discourse and Society* 11, 4 (2000): 459–485; Horgen, Kirsty. "Three million IVF babies born worldwide." *BioNews,* http://www.bionews.org.uk/new.lasso?storyid=3086 (accessed April 2007); Inhorn, Marcia, and Frank van Balen, eds. *Infertility Around the Globe: New Thinking on Childnessness, Gender, and Reproductive Technologies.* Berkeley: University of California Press, 2002; Marsh, Margaret, and Wanda Ronner. *The Empty Cradle: Infertility in America from Colonial Times to the Present.* Baltimore, MD: The Johns Hopkins University Press, 1996; Pfeffer, Naomi. *The Stork and the Syringe: A Political History of Reproductive Medicine.* Cambridge: Polity Press, 1993; Saetnan, Anne, Nelly Oudshoorn, and Marta Kirejczyk. *Bodies of Technology: Women's Involvement with Reproductive Medicine.* Columbus: The Ohio State University Press, 2000; Spallone, Pat, and Deborah Steinberg. *Made To Order: The Myth of Reproductive and Genetic Progress.* Oxford: Pergamon Press, 1987; Throsby, Karen. *When IVF Fails: Feminism, Infertility and the Negotiation of Normality* Basingstoke: Palgrave, 2004; World Health Organization. "Infertility," http://www.who.int/topics/infertility/en.

Karen Throsby

History of Birth Control

Birth control—also known as contraception—refers to the numerous methods and tools used by women and men to prevent conception and pregnancy. Although many contemporary birth control methods involve scientific and medical advancements, contraception is not a new practice. As long as humans have understood the connection between sexual intercourse, conception, and pregnancy, men and women have tried to prevent unwanted births. People have been practicing contraception for millennia—along with abortion and infanticide—as a means to control the birth and spacing of children.

Four of the most common contemporary methods include medically prescribed hormones, barriers, behavioral techniques, and irreversible sterilization. Hormones such as progestin and estrogen are prescribed to women and act as a contraceptive by preventing ovulation (the release of an egg by the ovary that is part of a woman's menstrual cycle), fertilization (the uniting of sperm and egg), and in some cases implantation (the attachment of a fertilized egg to the uterus).

Barrier methods include condoms, diaphragms, and sponges. Condoms (including the female condom) prevent conception by collecting semen during intercourse. Diaphragms and sponges work by blocking the uterus and preventing sperm from reaching the egg. Barrier methods are sometimes combined with spermicides, which come in the form of foams, jellies, and creams, and work by immobilizing sperm so that they cannot reach or fertilize the egg.

Behavioral methods include abstinence, withdrawal before ejaculation, and fertility awareness, which involves avoiding intercourse during the fertile times of a woman's menstrual cycle. Irreversible sterilizations are also a birth control option: men may be permanently sterilized with a vasectomy, and women may be sterilized with a tubal ligation, popularly referred to as "tube-tying."

With this wide variety of birth control methods, one may think of contraception as a matter of consumer choice and preference. Decisions regarding birth and reproduction may also be framed in terms of ideas about procreative autonomy and bodily integrity, that is, individual rights and liberties to make choices about what happens to one's own body. Deciding when or if to have a child, or multiple children, obviously involves these individual and familial choices, but there are structural determinants that inform the very fields upon which those individual negotiations take place: religious and moral frameworks, laws and regulations, scientific and medical knowledge, and economics and standards of living.

Margaret Sanger. Courtesy of Library of Congress, LC-USZ62-105456.

Individual decisions add up to demographic aggregates, and the state, for example, has an interest in regulating its population. By permitting, prohibiting, or forcing a population or social group to practice birth control, the state can regulate it demographically at the same time it transmits norms about gender and sexuality. Patriarchal, religious, and medical authorities also have interests in procreative decision-making. These interests force us to think beyond the individual rights-bearer, or more specifically, the rights-bearing body. Bodies, particularly the bodies of women—because it is women's bodies that are capable of conceiving and bearing children—become the sites of power and resistance in battles over who ultimately has control over conception, pregnancy, and birth.

The history of birth control is marked by the tension between its dual nature as a potential mechanism for both emancipation and coercion. This tension is indicated in the very term "birth control," which was coined by Margaret Sanger (1879–1966), founder of the American Birth Control League and Planned Parenthood, in 1915. Although today we commonly use "birth control" interchangeably with "contraception," Sanger's term specifically reflected the feminist desire for individual women to exert agency in their reproductive choices.

Ancient, Folk, and Traditional Methods

Attempts to control the number and spacing of children have of course existed since long before Sanger coined the term. Ancient Egyptians, for example, used herbs as contraceptives and abortifacients thousands of years ago; their herbal concoctions included acacia and honey and were inserted as vaginal suppositories. Greeks and Romans used contraceptive plants and herbs such as pomegranate seeds, pennyroyal, Queen Anne's lace, and myrrh. Hebrew texts speak of pessaries (vaginal suppositories) made of wool and flax, and of roots ingested by men and women that cause sterility.

Cultures that used folk methods did not only depend on medicinal herbs and teas; they also relied on magic, barriers, and behaviors. Men and women throughout the ages have practiced magic rituals such as dances, prayers, wearing amulets, and tying knots. Women have inserted barriers made out of cloth, sponges, bamboo, moss, or seaweed in order to block the cervix. Islamic, Indian, and African societies may have used dung as a spermicidal pessary. Traditional behavioral methods throughout the world have included *coitus interruptus* (that is, withdrawal), fertility awareness, and douching. Behaviors have even included sneezing or jumping up and down after intercourse as a means to prevent conception.

Before the Christian era, birth control, while not necessarily encouraged, was also not prohibited among Greek, Roman, Egyptian, Hebrew, and West Asian peoples. By the Middle Ages, the Christian Church began to take a strict stand against birth control. Despite religious prohibitions, herbal contraceptives and abortifacients continued to be used in Europe throughout the Middle Ages and the early modern period. Birth control continued to be an accepted practice in many Arab societies.

In the New World, a mix of indigenous, European, and African folklore contributed to the contraceptive knowledge passed on by word of mouth, by midwives, and in medical almanacs. Coitus interruptus, herbally induced abortions, and continuous breastfeeding (which may delay ovulation) were the primary methods of limiting pregnancies in seventeenth- and eighteenth-century America. Condoms, though not yet widely used as a contraceptive, became available in the eighteenth century. By the nineteenth century, douching, spermicides, and fertility awareness were popular contraceptive methods in the United States.

The Birth Control Movement

Prior to the late nineteenth century, birth control in the West was largely a matter of private, whispered knowledge. One of the routes by which birth control entered the public sphere was through the quasi-feminist voluntary-motherhood movement, in which advocates argued that by practicing abstinence or coitus interruptus, women could gain more dignity rather than be subjected to the sexual will of their husbands. While some advocates linked voluntary motherhood with free love or the agitation for women's political rights of suffrage, another strain of the movement was linked with Victorian moral reform and the rejection of male sexuality and intemperance. Voluntary motherhood was largely a cause championed by virtuous bourgeois white ladies who did not question their primary role as mothers, but who did begin to question the terms on which they entered into and experienced motherhood.

That voluntary motherhood rejected condoms, diaphragms, and sponges in favor of behavioral methods reflected the stigma that associated contraceptive devices with sexual profligacy. Even more so than behavioral methods, contraceptive devices highlight the active, premeditated separation of intercourse from procreation. This separation raises moral questions about their use. The Catholic Church and some other religious institutions have long condemned contraception as a mortal sin that goes against biblical mandates, for example, to "be fruitful and multiply."

Despite the moral and religious restrictions that shaped reproductive decisions, *legal* prohibitions against contraception did not arise in the United States until 1873, with the passage of the anti-obscenity Comstock Act in 1873—a federal law along with similar prohibitions in 24 states, collectively know as the Comstock Laws. This law made it illegal to distribute "obscene" material through the mail, thus effectively banning the advertisement, promotion, and sale of contraceptives, as well as any information about abortion.

The Comstock Laws are one example of how states intervene in procreative decision-making as a means to control the population both demographically and in terms of sexual behavior. The Comstock Laws may be interpreted as a governmental response to changing sexual norms and demographic realities. The voluntary-motherhood movement threatened to upend not only the gendered social order, but because it was primarily a movement of affluent white women, it also threatened to upend the racial social order: the fewer children born to affluent white families, the greater the proportion of children born to immigrants, people of color, and the working class and poor of all races and ethnicities in the U.S. population. In addition to the Comstock Laws' ban on contraceptives, a propaganda effort in the late nineteenth and early twentieth centuries urged white bourgeois women to stop committing "race suicide" by practicing birth control. They were urged to uphold their republican duties by reproducing the nation and nurturing a new generation of citizens.

It was in this political atmosphere that the U.S. birth control movement emerged. In 1914, Sanger was charged with violating New York's Comstock Law when she publicly urged women to limit their pregnancies in her socialist journal *The Woman Rebel*. Sanger, who was trained and worked as a nurse, advocated with other activists to promote contraception in their publications, distribute contraceptives in birth-control clinics, lobby for its legalization, and, significantly, urge the medical establishment to develop more effective methods. The birth control movement described contraception as a "right" of women to decide if, when, and how many children to bear—a right that would be echoed in the abortion rights movement—without intervention from the state or religious institutions. Birth control was a matter of economic and gender justice in an age when Sanger and her colleagues were distressed by the stories of women they met who were burdened financially, politically, socially, and emotionally by unplanned children.

In both the voluntary motherhood and U.S. birth control movements we can clearly see the emancipatory potential of birth control: as the "second sex," women have been unduly burdened by their ability to conceive and birth children. Being "voluntary" rather than involuntary mothers gave women a measure of agency within their own homes and marriages, and demanding the right to use contraception gave women agency as social and political actors. At the same time, the coercive aspect to birth control was also embedded in the rhetoric of the birth control movement, as contraception was seen as a way to lessen the burden that increasing numbers of children had not only on individual families, but on society as a whole. The birth control movement in fact occurred simultaneously and was deeply intertwined with the eugenics movement of the late nineteenth and early twentieth centuries.

Sanger and other birth control advocates aimed to increase women's access to contraception not only as a tool of liberation, but also as a eugenic tool for "race betterment." Although Victorian voluntary mothers were criticized by eugenicists for committing "race suicide" by decreasing the numbers of the white bourgeoisie, eugenicists were even more dismayed by the so-called threat that increases in the populations of the poor and working-class, immigrants, people of color, and the mentally and physically impaired posed to the nation. Birth control advocates thus found allies in the eugenics movement, who heralded contraception, notably permanent sterilization, as a solution to social problems such as poverty, insanity, and criminality because it would ensure that indigent, mentally ill, and otherwise "undesirable" populations would not reproduce. Tens of thousands of women and men were compulsorily sterilized in North America and Europe.

Feminist and critical race scholars have chastised the early birth control movement's use of and alliance with the virulently racist ideology, discourse, and practice of eugenics, arguing that rather than liberating all women, contraception was used as a method of social control for the poor, people of color, immigrants, and those with disabilities. One way to parse this dual nature of birth control is to see the *voluntary* use of contraceptive devices by white middle-class women as being emancipatory at the same time that *compulsory* birth control (for example, sterilization) restricted the reproduction of those who were less privileged or publicly demonized.

The medicalization of birth control is another aspect wherein feminist scholars argue that the birth control movement may have actually limited rather than expanded the agency of women. Rather than keep birth control as a woman-centered practice outside of the medical establishment, Sanger herself urged women to visit doctors to be prescribed with and fitted for contraceptive devices, and sought to win support for contraception from the medical profession in order to grant legitimacy to the birth control movement. Feminist scholars are critical of Sanger for relinquishing power to manage birth control to physicians (who were usually male) rather than keep it in the hands of women. For example, as medical sociologist Barbara Katz Rothman (2000) writes, "Sanger and the birth-control movement made alliances that were probably necessary, but there are costs, and we are still paying them.... Fitting a diaphragm is no harder than fitting a shoe. Women can do it for themselves. Yet birth control was incorporated firmly into the practice of medicine—a diaphragm requires a prescription. Women coming for birth control information and services are 'patients'" (p. 75). Physician-prescribed contraception heralded a much different era than that of folkloric knowledge about herbs or magic rituals secretly passed down by word of mouth. Control of contraceptive information and technology may have become a matter of medical expertise and authority, but this very move also presaged scientific and medical investment in assessing contraceptive methods' reliabilities and effectiveness. Besides wanting the medical establishment to bestow legitimacy on birth control, throughout her career, Sanger repeatedly made entreaties to medical researchers to develop more effective methods; in particular, she advocated for a contraceptive that a woman could take as a pill. In the 1960s, Sanger's advocacy for both the legalization of birth control *and* a highly effective method would finally pay off.

Contraceptive Developments and Controversies

Women who desire to control whether or not they become pregnant or give birth have a wealth of contraceptives to choose from. These options vary in terms of reliability and effectiveness, and their use is circumscribed by access and availability, as well as legal and cultural restraints. Attitudes toward the use of contraception have shifted over time. Most states had overturned the Comstock Laws by the early to mid-twentieth century, but it was not until the 1965 Supreme Court decision in *Griswold v. Connecticut* that the use of contraceptives was finally legalized throughout the United States.

The *Griswold* case arose when the executive director and the medical director of the Planned Parenthood League of Connecticut were arrested and convicted for providing a married couple with information about and a prescription for contraception. The Supreme Court found that Connecticut's law against contraceptives violated the Fourteenth Amendment and the couple's rights to marital privacy. Thus contraception, as decided by the Supreme Court, was a matter that should be left to the individual couple, not the state. The *Griswold* decision was followed eight years later by the Supreme Court decision in *Roe v. Wade*, which legalized abortion in the United States on similar grounds in terms of individual privacy rights.

Griswold, and *Roe* after it, was won on liberal grounds regarding individual rights and liberties. The right to privacy (between a married couple or between an individual and her doctor) is not a right enshrined in the Constitution, but has been construed as either a right to private property (over one's body, one's bedroom, or one's marital contract), or to the pursuit of happiness. The grounds of both *Griswold* and *Roe* were far removed from the rhetoric of the voluntary-motherhood movement, in which women fought to free themselves from having to turn their bodies over to the mind of their husbands, or from the early birth control advocates, who fought to free women from having to turn their bodies over to the minds of lawmakers or religious institutions. Because the Supreme Court decided *Griswold* on the basis of *privacy*, questions regarding the personhood of women and rights to bodily integrity were left unsettled, and would remain so with *Roe*.

Post-*Griswold*, the story of birth control in the United States (aside from the parallel story of the struggle over abortion rights) is largely one of ongoing scientific and medical breakthroughs. Besides the overturning of the Comstock Laws, the 1960s saw another major development in terms of contraception—that is, the invention, marketing, distribution, and widespread use of the oral hormonal contraceptive popularly known as "the Pill." Indeed, popular demand for the Pill precipitated the *Griswold v. Connecticut* decision.

The Pill further cemented the role of medical intervention in separating intercourse from procreation: it is highly effective, and its timing is separated from the sexual act. The Pill also further shifted responsibility for contraception onto women, since hormonal contraceptives were not—and still have yet to be—developed for men. One could also argue, however, that the Pill actually gave women more control over contraceptive decisions, since they no longer had to negotiate with a male sexual partner. Furthermore, this highly effective contraceptive, which allowed for spontaneous sexual activity, felt liberating for many women in the age of the sexual revolution.

The 1990s and Beyond

One of the biggest contraceptive breakthroughs since the Pill's emergence in the 1960s was the development and marketing of long-term contraceptive solutions in the 1990s. Instead of ingesting pills on a daily basis, hormones could be implanted in the form of rods under the skin of a woman's arm in the case of Norplant, or injected right into her bloodstream as Depo-Provera. These long-term contraceptives have certain advantages and disadvantages compared to the Pill. Like the Pill, they are highly effective. Unlike the Pill, which a woman must remember to take every day, Norplant releases hormones continuously for two to five years, and Depo-Provera lasts for three months, thus both are less subject to user error. However, some critics have raised concerns about their side effects, which may be more severe than those of the Pill, and because they are long-term and cannot be reversed or stopped as easily. Others emphasize that long-term contraceptives are in effect temporary sterilization, and have the potential to be used as coercive or eugenic measures against marginalized populations such as poor women of color. This coercive potential is highlighted by the fact that implants and injections must be administered by medical staff and cannot be stopped or removed by the woman herself, unlike oral contraceptives, which a woman can stop ingesting if she is experiencing negative side effects or changes her mind about wanting to be on birth control.

The coercive and abusive potential of Norplant became apparent just days after its approval by the Federal Drug Administration (FDA) in 1990, when the newspaper *The Philadelphia Inquirer* printed an editorial suggesting that in order to solve poverty among African Americans, the government should offer financial incentives to women on welfare to be administered Norplant. Other instances of coercion also occurred in the 1990s, including several cases in which judges ordered female defendants pleading guilty to child abuse or drug use to be implanted with Norplant as a condition of parole, and legislators in dozens of states attempted to increase Norplant's availability to poor women with financial incentives and/or as a condition of receiving welfare benefits.

These cases and proposals have been likened to the involuntary sterilizations of the eugenics era, in that once again, marginalized groups of women were coerced or prevented from having children. These measures were also similarly racially targeted, since Norplant was most likely ordered to be given to black females convicted of child abuse or drug charges. The welfare proposals, while ostensibly aimed at all poor women on public assistance, were discursively linked to stereotypes of poor women of color, in particular so-called "welfare queens" who had children for increased welfare benefits. Although these particular proposals did not become legislation, some states in the mid-1990s did enact caps on the number of children that women on welfare could receive benefits for. Although initially marketed as a way to revolutionize contraception, both Norplant and Depo-Provera have been derided by some reproductive rights activists because of their coercive potentials, high costs, side effects, and a lack of informed consent in international clinical trials.

In the 2000s, two more contraceptive developments became available: a hormone-releasing patch worn on a woman's skin that must be changed every week, and a hormone-releasing ring that is inserted by the woman into her vagina once a month. These devices have been marketed to young women in particular, who are perceived as less likely to remember to take an oral contraceptive pill on a daily basis. Unlike implantable or injectable contraceptives, the patch's application is controlled by the user rather than a medical practitioner, and its usage is much more short-term. However, the FDA warns that the contraceptive patch may cause blood clots because of the higher levels of estrogen released compared to those in the Pill. These contraceptive developments notwithstanding, critics have argued that rather than continuing to invent and market new methods of hormonal contraception for women, scientists should prioritize developing male contraceptives, lessening the burden on women to be responsible for birth control.

Contraceptive Access and Constraint

Despite the fact that there are a number of contraceptive options, these choices are a reality only when women have access to them. Access may be limited by affordability, availability, legal restraints, and cultural norms. Although birth control has at times been put to sinister uses, women have long looked for ways to limit pregnancies and births. When contraception is available— whatever those methods are—women use it. According to the United States Centers for Disease Control and Prevention (CDC), between 1982 and 2002, 98 percent of U.S. women between the ages of fifteen and forty-four who have ever had sexual intercourse with a male partner have used at least one contraceptive method or device; 62 percent are currently practicing contraception. Use of contraceptives, however, varies by socioeconomic status, race and ethnicity, age, religion, education, and many other factors.

Women worldwide have demanded access to birth control, and the use of contraceptives varies globally. According to a 2005 report by the United Nations Department of Economic and Social Affairs, 60.5 percent of married women aged fifteen to forty-nine worldwide are currently practicing some form of contraception. Contraceptive use tends to be higher in more developed regions than in less developed regions. The outlier country, however, is China (including Hong Kong), with 86.2 percent of women using some method of contraception. This probably reflects the one-child policy imposed by the Chinese government, and thus serves as a reminder of the role that the state has in the reproductive decisions of its populace. Besides economics and government intervention, disparities in contraceptive use and access also reflect the role of cultural norms. Although contraceptive use tends to be less frequent in developing nations, prevalence has grown as birth control, family planning, and global development programs and non-governmental organizations (NGOs) have increased their efforts to augment access and availability.

Increasing worldwide access to contraception may be framed as a global human development issue, linking birth control with more power and agency for women. The United States Agency for International Development (USAID) advocates voluntary family planning for its health, economic, and social benefits on both local and global levels. USAID champions family planning for decreasing high-risk pregnancies, improving the health of women and children, fighting HIV/AIDS, reducing abortions, increasing women's access to jobs and education, and protecting the environment.

Yet the coercive potential of birth control may also be embedded in international family planning programs. The impetus for population control efforts lies in part in post-World War II moral panics—indicated by sensationalistic terms such as "population explosion" and "population bomb"—in which "overpopulation" in the developing world was blamed for using up the planet's scarce resources and for increasing pollution, poverty, and violence. Fearing that growing populations could upset geopolitical, financial, and ecological stability, some advocates have framed bringing contraceptives and family planning practices to developing countries as a global public health issue. Critics, however, contend that population control has the potential to become eugenics on a global scale.

The birth control movement itself shifted in its emphasis on accessible birth control for all to the ideologies of family planning and population control. Note that Margaret Sanger's organization was originally named the American Birth Control League but in 1942 became the Planned Parenthood Federation of America. In a letter to the *New York Times* published in 1960, Sanger argued that, "Birth control, family planning, and population limitation are most important in any effort to bring real peace in the world. Less population will bring less war. Fewer people means more peace. Population planning through Government health and welfare departments will both improve the health, welfare, and happiness of children, mothers, and families and provide the essential foundations of world peace." There is a marked shift in birth control being based on decisions of the individual woman in terms of family size, to that of decisions of the state in terms of demographics and peacekeeping—and thus a shift in the locus of control.

Family planning and birth control on both the local and global level highlight the interactions between state regulations and cultural norms about gender, sexuality, abortion, and childbearing that are continually being contested. For example, U.S. law not only forbids foreign family planning programs from using any U.S. funds in support of abortions, but as of the 2001 Mexico City Policy (also known as the "Global Gag Rule"), nongovernmental organizations must certify that they will not perform or promote abortions as a condition of receiving *any* USAID assistance for family planning.

Aid for family planning may be required to bring down costs, educate consumers, and provide the infrastructure for birth control distribution. However, even with a proper infrastructure, contraceptive access throughout the world is also affected by local norms about ideal family size, the sexual autonomy of women to negotiate birth control with their partners, and by the levels of trust that people have in international family planning organizations or medical practitioners.

The interplay between contraceptive access and cultural values around gender, sexuality, and childbearing is not only an issue in developing nations of the global south, but is also evident in wealthy nations such as the United States, which has experienced controversy over emergency contraception (also known as Plan B or the "morning-after pill"). In August 2006, the FDA approved Plan B—an oral contraceptive pill that is taken after unprotected sexual intercourse to prevent pregnancy—for over-the-counter sales for women eighteen years of age or older by licensed pharmacists. Women seventeen or younger may obtain Plan B only by prescription.

The FDA's qualified approval may be seen as a compromise between those seeking the ready availability of emergency contraception, and religious and conservative figures who condemn emergency contraception for promoting sexual activity or even deride it as a form of abortion. This controversy harkens back to debates about the morality of contraception, and the ways in which legal, religious, and cultural structures impact the availability of birth control. It also highlights the ways in which birth control and abortion continue to be entangled and enmeshed in the public and political consciousness.

The history of birth control encompasses several interlocking and overlapping stories. One is about knowledge and practice, as women and men throughout the world have made many attempts—using a wide variety of herbs, drugs, behaviors, rituals, and devices—to prevent conception, pregnancy, and childbirth. These contraceptive methods sometimes have been developed in laboratories and prescribed by physicians, but they have also included plants and herbs gathered from the earth and sea, and by way of folkloric knowledge passed down by word of mouth and from generation to generation.

A second story of birth control involves its politics, that is, the push-pull dynamic among multiple forces—individual freedom and agency, government intervention, religious morality, medical authority, and scientific research—exercised upon and struggled over the bodies of women. This story also involves knowledge and practice, in that it describes who has access to those very contraceptive methods and devices, to what end contraception should be used, and which parties retain interest in gaining control over reproductive processes. The Comstock Laws involved preventing the spread of contraceptive *knowledge*, and thus the spread of contraceptive *practice*, and the early birth control movement was a direct response to this.

The third story of birth control is its dual character as a tool of emancipation and of coercion. On the one hand, having the ability to decide whether or not to have a child is a liberating concept, an exercise of one's rights and liberties as a democratic subject. On the other hand, birth control has sometimes been construed as a moral obligation—or worse, as a punishment or even genocidal instrument—for those deemed unfit to reproduce. Contraceptive knowledge and practice yields stratified consequences for people from different social categories.

These three interwoven stories continue to play out all over the globe, yielding a variety of responses and circumstances. The procreative decisions that one makes regarding conception, pregnancy, birth, and motherhood do not happen in a vacuum. They are shaped, constructed, and limited by knowledge about methods (and their side effects), as well as the practical matters of availability and accessibility, affordability, sexual agency within one's relationship, government regulation, political atmosphere, public opinion, and cultural (including religious) norms and values. Women and men have been practicing birth control for thousands of years, and will continue to do so. Contraceptive knowledge and practice—in the three forms cited above as well as in other ways not enumerated here—will also continue to evolve.

Further Reading: Brodie, Janet Farrell. *Contraception and Abortion in Nineteenth-Century America*. Ithaca, NY: Cornell University Press, 1994; Cornell, Drucilla. *The Imaginary Domain: Abortion, Pornography & Sexual Harassment*. New York: Routledge, 1995; Gordon, Linda. *Woman's Body, Woman's Right: Birth Control in America*. New York: Penguin, 1990; Mosher, William D., Gladys M. Martinez, Anjani Chandra, Joyce C. Abma, and Stephanie J. Willson. "Use of Contraception and Use of Family Planning Services in the United States: 1982–2002." *Advance Data from Vital and Health Statistics* 350. United States Centers for Disease Control and Prevention, 2004; Riddle, John M. *Eve's Herbs: A History of Contraception and Abortion in the West*. Cambridge, MA: Harvard University Press, 1997; Roberts, Dorothy. *Killing the Black Body: Race, Reproduction, and the Meaning of Liberty*. New York: Pantheon Books, 1997; Rothman, Barbara Katz. *Recreating Motherhood*. New Brunswick, NJ: Rutgers University Press, 2000; Sanger, Margaret. "Population Planning: Program of Birth Control Viewed as Contributing to World Peace." *New York Times*, January 3, 1960, E8; United Nations Department of Economic and Social Affairs, Population Division. "World Contraceptive Use 2005," http://www.un.org/esa/population/publications/contraceptive2005/WCU2005.htm (accessed July 27, 2006).

Lauren Jade Martin

History of Childbirth in the United States

The history of pregnancy and birthing is the history of midwifery, and the recorded history of midwifery goes back as far as recorded history. The wife of Pericles and the mother of Socrates were midwives, and in the book of Exodus, Moses' midwives, Shifrah and Puah, were rewarded by God. Every culture in the history of the world has had its midwives; wherever there have been women, there have been midwives.

Early America was no exception. Midwives settled as part of the Pilgrim population, and they attended births at the home of the birthing woman along with her female friends and relatives. The eventual downfall of the midwife in the United States was anticipated in European and British practice. In England, under the guild system that developed in the thirteenth century, the right to use surgical instruments belonged officially to the surgeon. Thus, when giving birth was absolutely impossible, the midwife called in the barber-surgeon to perform an embryotomy (crushing the fetal skull, dismembering it in utero, and removing it piecemeal), or to remove the baby by cesarean section after the death of the mother. However, it was not within the technology of the barber-surgeon to deliver a live baby from a live mother.

Not until the development of obstetrical forceps, by the barber-surgeon Peter Chamberlen (1560–1631) in the early seventeenth century, were men involved in live births. When the Chamberlens, after keeping the family secret for three generations, finally sold their design (or it leaked out), it was for the use of barber-surgeons and was not generally available to midwives.

It has frequently been assumed that the forceps were an enormous breakthrough in improving maternity care, but on careful reflection that seems unlikely. Physicians and surgeons did not have the opportunity to observe and learn the rudiments of normal birth and were therefore at a decided disadvantage in handling difficult births. And, whereas in pre-forceps days, a barber-surgeon was summoned only if all hope of a live birth was gone, midwives were increasingly encouraged and instructed to call in the barber-surgeon prophylactically, whenever birth became difficult.

In the eighteenth century, male birth attendants became increasingly more common, and having a male birth attendant became a mark of higher social status. Concurrently, spurred on by the development of basic anatomical knowledge and increasing understanding of the process of reproduction, surgeons began to develop formal training programs in midwifery. Women midwives were systematically excluded from such programs. While some men surgeons did try to provide training for midwives, sharing with them advances in medical knowledge, such attempts failed in the face of opposition from within medicine, supported by the prevailing beliefs about women's abilities to perform in a professional capacity. The result was a widening disparity between midwives and surgeons.

With a monopoly over tools developed to assist with difficult births, physicians came to be socially defined as having expertise in the management of these challenging or abnormal births, and so midwifery effectively lost control over even normal birth. The surgeon was held to be necessary "in case something goes wrong," and the midwife became dependent on the physician and his goodwill for her "backup" services. When physicians want to compete for clients, all they have to do is refuse to come to the aid of a midwife who calls for medical assistance. This is a pattern that began in the earliest days of the barber-surgeon and continues today.

In the nineteenth and early twentieth centuries, midwives and physicians in the United States were in direct competition for patients, and not only for their fees. As newer, more clinically oriented medical training demanded "teaching material," doctors sought to acquire as many patients as possible, turning to the previously undesired populations of poor and immigrant women. Doctors used everything in their power to stop midwives from practicing. From spreading false and racist advertising to utilizing political sway, physicians drew patients in, and, state by state, restricted midwives' sphere of activity and imposed legal sanctions against them.

The success of the physicians' campaign was evident in Washington, D.C., where the percentage of births reported to be attended by midwives shrank from 50 percent in 1903 to 15 percent in 1912. However, infant mortality in the first day, first week, and first month of life all increased. New York's dwindling corps of midwives did significantly better than did New York's doctors in preventing both stillborns and puerperal sepsis (postpartum infection). In Newark, New Jersey, a midwifery program in 1914–1916 achieved maternal mortality rates as low as 1.7 per thousand, while in Boston, in many ways a comparable city but one where midwives were banned, the rate was 6.5 per thousand.

Nevertheless, medicine gained virtually complete control of childbirth in the United States. Midwifery almost ceased to exist, and for the first time in history, an entire society of women was attended in childbirth by men.

The standards for obstetrical intervention that gained acceptance in the 1920s and 1930s remained in place through the 1970s, and shadows of those practices remain today. These practices can be traced back to a 1920 article in the *American Journal of Obstetrics and Gynecology*, "The Prophylactic Forceps Operation," by Joseph B. DeLee (1869–1942) of Chicago. DeLee's procedure for a routine, normal birth required sedating the woman through labor and giving ether during the descent of the fetus. The baby was to be removed from the unconscious mother by forceps. An incision through the skin and muscle of the perineum, called an episiotomy, was to be done before the forceps were applied. Removal of the placenta was also to be obstetrically managed rather than spontaneous. Ergot, a fungus found on grains and grasses, or a derivative was to be injected to cause the uterus to clamp down and prevent postpartum hemorrhage. The use of forceps was to spare the baby's head, DeLee having famously compared labor to a baby's head being crushed in a door. But the use of forceps also required episiotomies. The implication of the DeLee approach to birth for the mother is that she experienced the birth as an entirely medical event, not unlike any other surgical procedure.

Focus on anesthesia continued, and in the 1930s, 1940s, and 1950s, the prevalent form of sedation was "twilight sleep," a combination of morphine for pain relief in early labor, and the amnesiac, scopolamine. Women in twilight sleep therefore had to be restrained lest their uncontrolled thrashing cause severe injuries, as the drugs caused a potentially terrifying combination of disorientation and pain. The birth itself was not part of the mother's conscious experience because she was made totally unconscious for the delivery. These women required careful observation as they recovered from anesthesia; it might be quite some time before they were told the birth was over. The babies themselves were born drugged and required careful medical attention. The drugged, comatose newborn was the source of popular imagery of the doctor slapping the bottom of the dangling baby, attempting to bring it around enough to breathe. It could be several hours, or even days, before the mother and baby were "introduced."

While some women wanted pain relief and supported doctors in the development of twilight sleep, so-called "painless" labor, and medicalization, other women fought this turn of events. There were instances of American women rejecting one or another aspect of modern obstetrics long before any "movement" began. Anthropologist Margaret Mead (1901–1978), for example, wrote in her autobiography, *Blackberry Winter* (1978), of her attempts to recreate for herself in a U.S. hospital the kind of labor she had seen so often in the South Seas.

The first major thrust of the childbirth movement, however, was the publication of *Natural Childbirth* by Grantley Dick-Read (1890–1959), an English obstetrician, in 1933 (U.S. edition, 1944). Dick-Read developed his concept of "natural childbirth" as a result of a home birth experience. As the baby's head was crowning (beginning to emerge), he offered the mother a chloroform mask, which she rejected. She went on to have a calm, peaceful, quiet birth. When Dick-Read asked her later why she had refused medication, she gave an answer that became a classic statement of the natural childbirth philosophy: "It didn't hurt. It wasn't supposed to, was it, doctor?" (Dick-Read, 1944: 2).

In the years that followed, Dick-Read decided that no, it was not "meant to hurt," and developed a theory of natural childbirth. In a nutshell: "If fear can be eliminated, pain will cease." While Dick-Read's book and method appealed to many American women, women in the United States who attempted to follow his advice did so under hostile conditions. They were confined to labor beds, they shared rooms with women who were under scopolamine, and their screams, combined with the repeated offering of pain-relief medication by the hospital staff, reinforced the very fear of birth that Dick-Read set out to remove.

What was needed was a childbirth method designed to meet the needs of hospitalized women in the United States. The Lamaze or psychoprophylactic method met those demands. Lamaze grew out of Pavlovian conditioning techniques, originating in the Soviet Union where it was the official method of childbirth, observed by Fernand Lamaze (1891–1957) and Pierre Vellay, French obstetricians, and brought to France. In 1956, Lamaze published a book on the method, *Painless Childbirth*, and Pope Pius XII sanctioned its use. Worldwide acceptance followed swiftly. The method was brought to the United States not by an obstetrician, but by a mother, Marjorie Karmel, an American who gave birth with Lamaze in Paris in 1955. In 1959, she published *Thank You, Dr. Lamaze: A Mother's Experience in Painless Childbirth*. In 1960, Karmel, along with physiotherapist Elizabeth Bing and physician Benjamin Segal, founded the American Society for Psychoprophylaxis in Obstetrics (ASPO).

In the Lamaze technique, uterine contractions were held to be stimuli to which alternative responses could be learned; it was believed that with training, the pain and fear responses most women had learned could be unlearned, or deconditioned, and replaced with such responses as breathing techniques and abdominal "effleurage," or stroking. Lamaze technique introduced the rhythmic "puffing and panting" that characterizes, or maybe caricatures, the portrayal of birth in modern media. Concentration on these techniques—based on the methods used by midwives—inhibited cerebral cortex response to other potentially painful stimuli.

ASPO, unlike the Dick-Read approach, was specifically geared to U.S. hospitals and the American way of birth. Rather than challenging the authority of obstetrics and the practices that obstetricians employed, ASPO encouraged women to:

> ... respect her own doctor's word as final ... It is most important to stress that her job and his are completely separate. He is responsible for her physical well-being and that of her baby. She is responsible for controlling herself and her behavior. (1961: 33)

The physical trappings and procedures surrounding U.S. births, including perineal shaves, enemas, delivery tables (although women were taught to request politely that only leg and not hand restraints be used), episiotomies, and so on, were not questioned by ASPO. Even more basically, ASPO accepted the medical model's separation of childbirth from the rest of the maternity experience, stating in this first manual that rooming-in (mother and baby not being separated) and breastfeeding are "entirely separate questions" from the Lamaze method.

Contemporary Pregnancy Experience

The modern birthing experience raises contradiction after contradiction. The midwife has returned and there has been an increase in the cesarean section. For fifty years, there were almost no midwives to be found in most of the United States. Then, during the same period when nurse-midwives began practicing in U.S. hospitals, while the "natural childbirth" movement flourished and hospitals introduced "birthing rooms" to replace operating room-like delivery rooms, the cesarean section rate rose from slightly over 5 percent in 1970 to an astonishing and unprecedented 29.1 percent in 2004; it increased by 41 percent between 1996 and 2004.

Since the development of modern obstetrics, there has never been more talk of birth as a "healthy, natural event," yet each individual birthing woman is now acquainted with her personal "risk factors." Nearly all U.S. women give birth in hospitals, and of these, 85 percent are strapped to fetal monitors, despite evidence that such monitoring produces unnecessary interventions.

The current vocabulary of "risk" replaced the vocabulary of "disease." All pregnant women are seen as at risk of developing any number of conditions. Thus, the medical model of prenatal care is essentially a process of screening for pathology. More and more women are defined as having "high-risk" pregnancies as physicians claim an ever-expanding region of pregnancy as their own. Today, more pregnancies are defined as "high risk" than as "low risk."

The medical literature defines childbirth as a three-stage physiological process. In the first stage, "labor," the cervical opening of the uterus into the vagina, dilates from being nearly closed to its fullest dimension of approximately ten centimeters (almost four inches). In the second stage, the baby is pushed through the opened cervix and through the vagina, or birth canal, and out of the mother's body. This is "delivery." The third stage is the expulsion of the placenta or "afterbirth."

Williams Obstetrics, the primary medical reference in obstetrics, states: "One of the most critical diagnoses in obstetrics is the accurate diagnosis of labor" (Cunningham et al., 2001:310). They offer two reasons: "If labor is falsely diagnosed, inappropriate interventions to augment labor may be made" (310). The second reason *Williams* offers is that if labor is not diagnosed, "the fetus-infant may be damaged by unexpected complications occurring in sites remote from medical personnel and adequate medical facilities" (310). The birth may, that is, occur at home. Once at the hospital, though, physicians seek to manage labor "proactively," actually taking charge. But, from the physician's perspective, there is not much that can be done to manage labor. Doctors cannot make, produce, or guarantee healthy babies or healthy mothers. The two areas they have had some success with are the management of the pain of labor and its length.

The epidural marks recent developments in the medical management of the labor. According to the Maternity Center Association's report *Listening to Mothers* (2006), 59 percent of women having vaginal births reported using epidural analgesia for pain relief during labor. For an epidural, a fine plastic tube is inserted in the lower back just outside the spinal cord, and numbing drugs are dripped in. An anesthesiologist is necessary to insert the epidural, and the availability of on-site anesthesiologists is the single most important factor in a hospital's epidural rates. Epidurals have the enormous advantage of allowing the woman to be fully present mentally: she can observe the proceedings. When they work—and 15 percent of women do not get full pain relief from the epidural—they numb her.

Despite continuing focus on pain management, energy has shifted to the medical management of the length of labor. As long as one is prepared to stop the labor with a cesarean section, one can guarantee a labor no longer than any arbitrary time limit one chooses: twelve hours seems to have become the standard figure. In the United States today, according to the *Listening to Mothers* survey, almost half of mothers had a caregiver try to induce labor, and these inductions successfully started one-third of all labors. Just over half of these women reported that they were given drugs to strengthen (speed up) contractions, and more than half had their membranes artificially ruptured, also to speed up labor.

The standard obstetrical model of labor is best represented by "Friedman's curve." Obstetrician Emanuel Friedman observed labors and computed the average length of time they took. He broke labor into separate "phases" and found the average length of each phase. He did this separately for primiparas (women having first births) and for multiparas (women with previous births). He computed the averages and the statistical limits—a measure of the amount of variation.

What Friedman did was to make a connection between statistical normality and physiological normality. He used the language of statistics, with its specific technical meanings, to make conclusions about physiology: "It is clear that cases where the phase-durations fall outside of these (statistical) limits are probably abnormal in some way. We can see now how, with very little effort, we have been able to describe with proper degrees of certainty the limits of normal" (1959: 97). Once the connection is made between statistical abnormality and physiological abnormality, the door is opened for medical treatment. Statistically abnormal labors are medically treated. Treatments include rupturing membranes, administering hormones, and cesarean sections. Cesarean sections alone account for nearly 30 percent of births in the U.S. today.

"Doing something" is the cornerstone of medical management. Interventions, therefore, are the norm in the medicalized birth, whether intended to alleviate pain or limit the length of labor. With the medical focus on developing new technology and a seemingly limitless desire to screen mothers and fetuses as early as possible for as many abnormalities as possible, one can expect that the trend of medically managing birth and pregnancy will continue into the future with all of the force it exhibits today.

Further Reading: Bing, Elizabeth, and Marjorie Karmel. *A Practical Training Course for the Psychoprophylactic Method of Painless Childbirth.* New York: ASPO, 1961; Brack, Datha Clapper. "Displaced: The Midwife by the Male Physician." *Women and Health* 1 (1976): 18–24; Cunningham, F. Gary, Norman F. Gant, Kenneth J. Leveno, and Larry C. Gilstrap, 2001. *Williams Obstetrics*, 21st Edition. New York: McGraw Hill; Curtin, Sally C., and Lola Jean Kozak. "Decline in the U.S. Cesarean Delivery Rate Appears to Stall." *Birth* 25 (December 1998): 259–262; Friedman, Emanuel. "Cervitremy: a Study of the Parturient Cervix Uteri with Particular Reference to Dilation," doctoral thesis, Columbia-Presbyterian Medical Center. (1959); Goer, Henci. *Obstetric Myths versus Research Realities: A Guide to the Medical Literature.* Westport, CT: Bergin and Garvey, 1995; Guttmacher, Alan. *Pregnancy and Birth: A Book for Expectant Parents.* New York: New American Library, 1962.

Barbara Katz Rothman and Bonnie French

Menstruation-Related Practices and Products

Throughout history, women have developed methods to address the process of menstruation that have reflected the prevailing norms, values, and products of the times. One historical cultural attitude toward menstruation is that menstrual blood itself is dirty, contaminated, or even dangerous, and that all should avoid it. Another Western cultural attitude, particularly for women, is inconvenience, in that the flow of blood is thought to disrupt sexual activities, working, participating in sports, and wearing certain kinds of clothing. These outlooks have influenced how women manage menstrual blood each month.

Today in the developed world, most women use manufactured disposable tampons or menstrual pads to control the flow of their menstrual blood. Menstrual pads were the first to be commercially developed, and today have adhesive backings that attach to women's undergarments. Tampons absorb the blood from their position inside the vagina. These products are used because of their convenience, availability, and effectiveness in allowing women to conduct their normal physical activities during menstruation as well as protecting their clothing and maintaining a culturally desired level of hygiene. In the United States, approximately 70 percent of American women use tampons, with the vast majority of women preferring to use an applicator to insert a tampon, as compared with most European women, who prefer to use tampons without an applicator, known as a digital tampon. In the developing world, the use of manufactured pads and tampons during menstruation is influenced by cost, availability, and the cultural regard of women in general. For example, in some cultures, tampons are not available or widely used because they are thought to rupture a woman's hymen, which is an important form of proof of virginity. Another factor that influences the choice of products used to control menstrual blood flow is political ideology. Some women who are concerned about the environmental impact of disposable pads and tampons have sought alternatives, including reusable cloth pads, reusable tampons made from sea sponges or soft foam, and reusable menstrual cups made of plastic (brand names include the Keeper, Lunette, and the Moon Cup) that are placed in the vagina near the cervix in order to capture, rather than absorb, the blood. The most recent development in the process of controlling the menstruation process is the marketing of the injectable contraceptive Depo-Provera to totally suppress monthly periods, and the birth control pill Seasonale, which reduces a woman's periods to four per year.

Predecessors to Tampons, Pads, and Other Methods

Before the development of manufactured products like pads and tampons, women used a variety of available methods to address menstruation. Cloth or rags to absorb blood (which were washed and reused) were commonly used since at least the tenth century CE. American women during the nineteenth century would fashion a flannel or cloth diaper that would fit between their legs and be washed and reused. Women have used available absorbent materials in many cultures, including soft papyrus tampons by Egyptian women in the fifteenth century BCE, and wool inserted like a tampon in other societies. Women in ancient Japan were said to have made a type of tampon out of paper that was held in place with a bandage and changed up to a dozen times in a day. Traditionally, Hawaiian women have used absorbent native plants and grasses; mosses and other plants are still used by women in parts of Asia and Africa. For thousands of years, women also have used natural sponges to absorb menstrual blood.

This process of containing blood flow has continued through the centuries. The first menstrual cups designed to catch blood were said to have been manufactured as World War II began; however, their production was stalled because of the shortage of rubber. These cups were also not successful because they were hard and too heavy. Softer rubber cups that were manufactured as a mass-market product also did not sell well in the United States, probably because they need to be inserted into the vagina manually, and were considered unhygienic.

History of the Tampon and Pad

The first commercial menstrual pad in the United States was known by the brand name Lister's Towels, which were first manufactured in 1896 by Johnson & Johnson. However, these were not a commercial success and were made only until the mid-1920s. The first successful disposable menstrual pad was inspired by French nurses during World War I. They discovered that the wood-pulp material used for bandages made by the company Kimberly-Clark could absorb menstrual blood better than the cloth rags that they had been using, and that these bandages could be thrown away after each use. By 1921, Kimberly-Clark was manufacturing disposable menstrual pads named Kotex, a brand that continues to be sold today. All early sanitary napkins required the use of a separate belt for it to stay in place. By 1970, Personal Products Company introduced the brand Stayfree, and Kimberly-Clark released New Freedom pads, both featuring an adhesive backing that made belts unnecessary.

Tampons (from the French word Tampion which means stopper, plug, or buffer) that were manufactured for commercial use were available by the early 1930s, but none of these products (with such names as Fax, Nappons, Fibs, and Slim-pax) were accepted by the mainstream. None of these original tampons used an applicator to insert them, and this is thought to have had an effect on their popularity. It was not until Tampax tampons, which came with an applicator, were introduced in 1936 that tampons became popular. However, even Tampax encountered resistance from the public and doctors when they first appeared, as they were associated with showgirls, performers, and other women who were seen to have questionable values. Since tampon users were able to keep menstrual blood from flowing outside of their bodies, they were seen as being able to manage their sexually related bodily functions. Playtex introduced the first deodorant tampon in 1971, and in 1972, the National Association of Broadcasters lifted its ban on television advertising of sanitary napkins, tampons, and douches.

Toxic shock syndrome, an illness caused by bacteria and related to the use of super-absorbent tampons, became a public health issue in the 1970s. Fifty percent of cases were associated with Proctor and Gamble's Rely tampon, and the brand was taken off the market in 1981 (Brody 1983).

Further Reading: Brody, Jane E. "Scientists Unraveling Mystery of Toxic Shock," *New York Times,* April 26, 1983.

Martine Hackett

Semen

Cultural History of Semen

In the contemporary Western world, there are distinct differences between sperm, semen, and ejaculation. Biologically, semen is understood as a mixture of prostaglandin, fructose, and fatty acids. Sperm, or more accurately sperm cells, make up only 2 to 5 percent of a given sample of semen. Semen, along with its sperm, is released through the penis, commonly through ejaculation. In its natural state, an individual sperm cell exists alongside millions of others in semen. Yet, sperm as individual cells are commonly separated out from semen and anthropomorphized, that is, given human qualities, in a variety of contexts—in philosophy, children's facts-of-life books, and forensics and television crime shows, for example.

For centuries, scientists and scholars have attempted to define, manipulate, and control semen and sperm. Their work has been consistently informed and influenced by their own gendered and cultural understandings of men and masculinity, and even at times has been a direct response to various crises in masculinity. The scientific practices that elevate sperm to the lofty heights of sole reproductive actor also, ironically, transform sperm into a relatively easily accessible commodity. Sperm can be purchased and exchanged for profit. Modern reproductive technologies and services can now limit men's role in human reproduction to that of an anonymous sperm donor.

Sacred Sperm

The medieval Italian Catholic philosopher, St. Thomas Aquinas (1225–1274), wrote prolifically about the importance of seminal fluid and the morality of its ejaculation. According to Aquinas, who was greatly influenced by Aristotle, semen's intention is to produce a replica of itself, a male. Apparently, though, if the semen is "weak" or environmental factors are not precipitous, a female might get created instead. Semen was empowered by two sources. First, semen contained and was propelled by man's life force, his spirit, and his soul. Second, semen was interconnected with the forces of the universe. Aquinas wrote in *Summa Theologica II* in 1273:

Donor siblings during their reunion in 2006 in Fresno, California. The three children were conceived from the same sperm donor and their mothers located each other through the Donor Sibling Registry website. (AP Photo/Gary Kazanjian)

> This active force which is in the semen, and which is derived from the soul of the generator, is, as it were, a certain movement of this soul itself: nor is it the soul or a part of the soul, save virtually; thus the form of a bed is not in the saw or the axe, but a certain movement towards that form. Consequently there is no need for this active force to have an actual organ; but it is based on the (vital) spirit in the semen which is frothy, as is attested by its whiteness. In which spirit, moreover, there is a certain heat derived from the power of the heavenly bodies, by virtue of which the inferior bodies also act towards the production of the species (as stated above). And since in this (vital) spirit the power of the soul is concurrent with the power of a heavenly body, it has been said that "man and the sun generate man." (q. 18, art. 1, ad 3)

Further, the term for masturbation, "onanism," originating from the story of Onan in the book of Genesis. God killed Onan because during sexual intercourse with his dead brother's wife, Onan withdrew his penis, spilling his seed upon the ground.

The belief that semen is sacred continued to resonate in religious and cultural contexts into the seventeenth century, when some of the first scientific ideas about the form and function of sperm were developed. At this time, the theory of preformation asserted that within each primordial organism resided a miniature, but fully developed, organism of the same species. There were two competing theories of preformation: spermism versus ovism. Battles between the ovists, those who believe the preformed entity is in the egg, and spermists, those who believe the preformed entity is in the sperm cell, depict the different scientific constructions of how these cells embodied individual replicas of humans.

Most consider the spermists the victors in the battle. This culminated in the seventeenth-century notion that women were "mere vessels" in the context of human reproduction. But the consequences were also significant for men when, by the nineteenth century, beliefs about sperm's value centered on what has been called the "spermatic economy": seminal fluid is a limited resource that, once spent, cannot be recovered; therefore men should reserve ejaculation for vaginal sex with women. Further, because semen was believed to embody men's vital life forces, depleting one's supply on non-reproductive activities could be perceived as irresponsible at best and physically and/or morally weakening at worst.

Of course, official scientific knowledge about semen is produced at the same time as unofficial knowledge about semen is discovered and shared. In the case of sperm, folk ideas, remedies, and understandings of reproduction clearly indicate understanding of male bodily fluids. Over many centuries and across diverse cultures, women have attempted to block male fluid from entering their vaginas. There is evidence dating back to 1850 BCE Egypt that women inserted a pessary, or a soft ball containing various presumably spermicidal substances, into the vagina prior to intercourse. Prescientific medical knowledge was also available as early as 1550, when Bartholomeus Eustacus, an Italian homeopath, recommended that a husband guide his semen toward his wife's cervix with his finger in order to enhance the couple's chances of conception. This reference was the first recorded suggestion in Western medical literature that humans could control their reproductive capabilities by manipulating semen.

Scientists See Sperm

Throughout history, scientists have marveled at the sperm cells' powerful agency and self-contained role in reproduction. For example, in the 1670s, a single scientific innovation—the invention of the microscope—revolutionized the natural sciences, and provided an entirely new perspective for scientists studying semen. Antoni van Leeuwenhoek (1632–1723) was for the first time able to define sperm cells based on physical observation. Leeuwenhoek's own description of spermatozoa (sperm animals), or animalcules as he called them, perhaps indicates his excitement about this breakthrough. He marvels at the effort spermatozoa make to move minuscule spaces:

> Immediately after ejaculation ... I have seen so great a number that ... I judged a million of them would not equal in size a large grain of sand. They were furnished with a thin tale, about 5 or 6 times as long as the body, and very transparent and with the thickness of about one twenty-fifth that of the body. They moved forwards owning the motion to their tails like that of a snake or an eel swimming in water; but in the somewhat thicker substance they would lash their tails at least 8 or 10 times before they could advance a hair's breadth. (quoted in Kempers, 1976, 630)

While the use of a microscope allowed scientists to gain incredible insight into the physical aspects of previously "invisible" organisms, they could still only theorize about sperm's exact role in reproduction. For example, building on the work of Leeuwenhoek, researcher Nicholas Hartsoeker

(1626–1725) in 1694 theorized that the umbilical cord was born from the tail of the spermatozoa and the fetus sprouted from the head of a sperm cell embedded in the uterus. During the late 1700s, Lazzaro Spallanzani (1729–1799), an Italian priest and experimental physiologist, conducted research on fertilization using filter paper to investigate the properties of semen. He believed that spermatic "animals" were parasites and thought that the seminal fluid, not the sperm, was responsible for fertilization of the ovum. It was not until 1824 that French pioneers in reproductive anatomy, Pierre Prévost (1751–1839), a theologian-turned-naturalist, and Jean-Baptiste Dumas (1800–1884), a devout Catholic and chemist, repeated Spallanzani's filtration experiments. These scientists claimed that animalcules or sperm cells in seminal fluid fertilized female eggs. By the beginning of the twentieth century, two zoologists at the University of Chicago, Jacques Loeb (1859–1924) and Frank Lillie (1870–1947), established that sperm was species-specific and mapped out the physiology of fertilization.

As in the case of Leeuwenhoek's narrative above, scientists' gendered understandings of sperm and semen sometimes "leaks through" their scientific observations. For example, the well-known reproductive researcher, Robert Latou Dickinson (1861–1950), in his *Human Sex Anatomy* published in 1949, described semen as "a fluid that is grayish white rather than milky; upon ejaculation its consistency resembles a mucilage or thin jelly which liquefies somewhat within three minutes after emission, later becoming sticky" (1949: 81).

The Fertility Factor

In the twentieth century, many innovations in sperm research emerged out of scientists' attempts to understand male infertility. Men exposed to both environmental and occupational toxins reported consistently higher rates of infertility, yet there is no universal agreement as to its cause. There has been much debate in the fields of epidemiology, toxicology, and infertility regarding this increasing rate of men's infertility. As male fertility continues to be compromised by what some presume to be increasing environmental toxins, patients and researchers will more assiduously pursue biomedical solutions.

With such a wide variety of risk factors, it's not surprising that men's fertility has become increasingly compromised. For example, some diseases lead to male infertility, including cystic fibrosis, sickle-cell anemia, and some sexually transmitted diseases. Environmental factors such as pesticides and herbicides (including estrogen-like chemicals), hydrocarbons (found in products like asphalt, crude oil, and roofing tar), heavy metals (used in some batteries, pigments, and plastics), and aromatic solvents (used in paint, varnish, stain, glue, and metal degreasers) have all been suspected of lowering sperm counts and damaging morphology. (As with many aspects of seminal diagnostics, the determination of normal morphology is rife with fractious debate. Because sperm morphology is determined through a visual assessment, it is highly subjective.) Furthermore, men and boys may lower their sperm count through tobacco chewing and smoking. Prenatal exposure to tobacco has been shown to lower sperm counts in male offspring. Excessive alcohol consumption, marijuana smoking, and obesity also affect sperm counts and sperm performance. Endurance bicycling has been shown to significantly alter sperm morphology. Researchers are calling for studies of humans to determine if storing cellular phones near the testicles in the front pocket of pants decreases semen quality.

Normalizing Sperm—and Men

From their humble beginnings in the 1700s to today, scientific representations of sperm have come a long way. What was once a science based only on subjective narratives, sketches, and visual observation is now a probabilistic system that understands men's fertility based on a wide variety of quantifiable parameters. The parameters of semen analysis have expanded and now include volume, pH, viscosity, sperm density, sperm motility, viability, and sperm morphology.

The first diagnostic tool to be used in the detection of male infertility was the "spermatozoa count" or sperm count, developed in 1929 by scientists D. Macomber and M. Sanders. Sperm count was used for decades as the primary tool in measuring male infertility. However, imaging technologies, such as electron microscopes and other devices, created new opportunities to produce scientific knowledge about sperm. For example, biochemists can measure the components of human seminal plasma; one textbook, Marilyn Marx Adelman and Eileen Cahill's *Atlas of Sperm Morphology* (1989), listed thirty-five different elements of semen and their referent anatomical production mechanisms.

Currently, the two most common assessments of sperm's relative health are motility and morphology testing. Sperm motility can be measured quantitatively by counting the number of inactive sperm and/or qualitatively by assessing the type of movements sperm make. Scientists' efforts to measure the distance and speed of sperm movement has been remarkably varied and numerous. Morphology describes the shape of the sperm cell. The first attempt to classify sperm cells' morphology was in the 1930s, when scientist G. L. Moench sketched out fifty variations in sperm morphology with names such as microsperm, megalosperm, puffball, and double neck. Today, scientists typically define the normal shape as a sperm cell as having an oval configuration with a smooth contour. The acrosome, a cap-like structure on the sperm's head containing enzymes that help penetration of the egg, is well-defined, comprising 40 to 70 percent of the head. Further, there cannot be any abnormalities of the neck, midpiece, or tail, and no cytoplasmic droplets of more than half of the size of the sperm head. Given this definition of normal shape, a semen sample is considered to have a normal morphology when at least 12 percent of sperm cells are of normal shape.

Honorable Death—Kamikaze Sperm

In the late 1960s and early 1970s, Geoff Parker (1944–), a scientist researching dungflies, articulated a theory about competition between ejaculates of different males for the fertilization of the egg. Sperm competition occurs when sperm from two or more males is in the female reproductive tract at the same time. His work in the field has earned him the title "the father of sperm competition" by fellow scientists. Theories of sperm competition are predominantly tested and studied by evolutionary biologists and zoologists researching insects, birds, and non-human animals. Scientists have documented sperm competition in mollusks, insects, and birds and, using this data, have inferred that sperm competition occurs in humans as well. Many scientists theorize that female sexual infidelity is the context for sperm competition.

Expanding on this research, Tim Birkhead, a British behavioral ecologist, published *Promiscuity: An Evolutionary History of Sperm Competition and Sexual Conflict* in 2000. The book primarily explores the role of sperm competition and sperm choice in animal sexuality without a great deal of extrapolation to human sexuality. Indeed, Birkhead claims that to overgeneralize outside of a particular species is intellectually dangerous and scientifically irresponsible. But evolutionary biologists Robin Baker and Mark Bellis, in their book *Human Sperm Competition: Copulation, Masturbation, and Infidelity* (1999), have done just that. They apply the scientific principles of sperm competition through the generous extrapolation of animal models and behavior to humans. Baker and Bellis claim that human sperm competition is one of the key forces that shapes genetics and drives human sexuality. Their metaphor for human sperm competition is strategic warfare, wherein sperm range from generals to enlisted soldiers, tragic heroes to victorious conquerors. Written in a casual and accessible style with many illustrations and diagrams, this work depicts the crossover between hard science and pop science.

Even if the science is questionable, the popularity of Baker and Bellis' book is understandable. Generally, in their ongoing attempts to figure out their place in the natural order, people derive great pleasure in having their beliefs about the origin of behavior proven through observations of animals. Donna Haraway, feminist biologist and political theorist, carefully examines the conventions of animal to human extrapolation. Haraway demonstrates how both scientists and laypeople have collaborated on using primates to understand human behavior, particularly the sexual nature of female primates as extrapolated to humans. Because these scientific observations about animals are viewed through our cultural mediations of race and gender socialization, what might be observed as supposedly naïve animal behavior is actually misperceived.

Critics of this approach point out that Baker and Bellis extrapolate liberally from bioscientific research primarily based on non-human models to grand theories of human behavior. Further, their uncritical promotion of sperm competition and its relevance for human behavior is devoid of sociological insights about the role of environment and socialization on humans. In addition, feminists point out that there are problematic statements that assume girls and women profit from sperm competition regardless of how sperm was placed in their bodies, whether consensually or nonconsensually.

The centerpiece of this book is the development of a theory and coining of the term the Kamikaze Sperm Hypothesis (KSH), which suggests that:

> animal ejaculates consist of different types of sperm each programmed to carry out a specific function. Some, often very few, are "egg getters," programmed to attempt to fertilize the female's eggs. The remainder, often the vast majority, is programmed for a kamikaze role. Instead of attempting to find and fertilize eggs themselves, their role is to reduce the chances that the egg will be fertilized by sperm from any other male. (23)

Through the KSH, Baker and Bellis provide a means to explain polymorphism of sperm and thus rescue all morphs of sperm from the label of useless, providing job descriptions for them all. "In the past, 'non-normal' sperm have been considered to be unwanted passengers in the ejaculate; unavoidable deformities that are a hindrance to conception. Our Kamikaze Sperm Hypothesis (KSH) argues otherwise. Each sperm morph has a part to play in the whole process of sperm competition and fertilization" (251). That is, each morph is a part of a team in pursuit of a larger goal of allowing the chosen sperm to fertilize the egg. In a battlefield motif, all sperm work in a concerted effort to benefit the chosen one's ability to capture the castle. Old and young sperm are "recruited" from the ejaculate: normal and non-normal morphs have purpose. And while "younger" sperm cells may be more likely to reach an egg, it is important to understand the ways in which the scientist delivers this information to us.

The authors discuss the credentials of these morphs of sperm in detail using football or war analogies to describe the "division of labor." For example, sperm cells labeled as the seek-and-destroy sperm engage in "head-to-head combat" (275), wherein the acrosomes on the sperms' heads are described as if they "were in effect carrying a bomb on their heads." Since sperm occupy two primary statuses, "egg-getters" and "Kamikaze," Baker and Bellis provide readers with detailed typologies of each. Egg-getters are thought to be "macros," sperm with oval-shaped heads that are longer and wider than other sperm. In addition to the normal challenges of sperm achieving fertilization, according to the KSH, egg-getters must overcome even more tremendous obstacles. "Throughout the whole journey, the successful egg-getter has to avoid the seek-and-destroy attention of both sperm from other males and even 'family planning' from some within its own ejaculate" (292).

Popularizing Human Sperm Competition

Some critics perceive such writings as guilty of a kind of antifeminist backlash historically expressed in scientific and medical discoveries and popularized through the mass media. As historian G. J. Barker-Benfield explained during the nineteenth century, the growth in women's rights "intensifies male anxieties" and greatly influences prevailing notions of men's, and, more extensively, women's reproductive capacities (53). Sociologist Michael Kimmel's research reveals "a strongly misogynist current runs through a number of religious tracts, medical treatises, and political pamphlets of the late nineteenth century. Opponents of economic, political, and social equality between men and women almost always resorted to arguments about the supposed natural order of things as counters to these social trends" (1997, 266). So while this practice of male anxiety over women's growth in social power is not new, in the case of sperm competition it takes on a new twist. In these cases, how sperm cells are characterized can tell us a great deal about contemporary attitudes toward men.

Theories of sperm competition are not just produced in the laboratory; they also circulate in popular culture, for example, the documentary *Enron: The Smartest Guys in the Room*, which chronicles the demise of the seventh-largest U.S. company into bankruptcy. Driven by the arrogance and greed of executives and the rapacity of Enron energy traders, the film illustrates the corporate culture created by Jeff Skilling, the CEO. The culture was one of intense masculine exhibitions (corporate retreats with mandatory motorcycle races on treacherous terrain) and bare-knuckled competition. As a means to dramatize the competitive environment of Skilling's design, the filmmakers employ cutaways to imagery of sperm magnified under microscopes. By his own admission, Skilling's favorite book, *The Selfish Gene*, influenced his belief that macho competition was predestined based on the winning quality of the genes (the sperm) that formed you (Dawkins, 1976).

Fluid Tenacity

Despite ongoing changes regarding the access to and means of human reproduction, sperm and semen maintain definitions of tenacity, strength, and supremacy. Much like the fluid of semen itself can leak onto different fabrics and into different bodies, the meanings of semen are able to seep into our consciousness, transmitting the validity of stereotypically conservative views of men and women.

Further Reading: Adelman, Marilyn Marx, and Eileen Cahill. *Atlas of Sperm Morphology*. Chicago: ASCP Press Image, 1989; Aquinas, Thomas. *Summa Theologica II*, q. 18, art. 1, ad 3 was written and published between 1265 and 1273; Baker, Robin. *Sperm Wars: The Science of Sex*. Diane Books Publishing Company, 1996; Baker, Robin, and Mark Bellis. *Human Sperm Competition: Copulation, Masturbation, and Infidelity*. New York: Springer, 1999; Barker-Benfield, Ben. "The Spermatic Economy: A Nineteenth-Century View of Sexuality." *Feminist Studies* 1:1 (1972): 45–74; Birkhead, Tim. *Promiscuity: An Evolutionary History of Sperm Competition and Sexual Conflict*. Cambridge, MA: Harvard University Press, 2000; Birkhead, Tim, and A. P. Møller, eds. *Sperm Competition and Sexual Selection*. San Diego: Academic Press, 1992; Check, J. H., H. G. Adelson, B. R. Schubert, and A. Bollendorf. "Evaluation of Sperm Morphology Using Ruger's Strict Criteria." *Archives of Andrology* 28:1 (1992): 15–17; Dawkins, Richard. *The Selfish Gene*. Oxford: Oxford University Press, 1976; Dickinson, Robert Latou. *Human Sex Anatomy*. Baltimore, MD: Williams and Wilkins Co., 1949; Farley, John. *Gametes and Spores: Ideas about Sexual Reproduction 1750–1914*. Baltimore, MD: The Johns Hopkins University Press, 1982; Haraway, Donna J. *Primate Visions: Gender, Race, and Nature in the World of Modern Science*. New York: Routledge, 1989; Kempers, Roger. "The Tricentennial of the Discovery of Sperm." *Fertility and Sterility* 27:5 (1976): 630–635; Kimmel, Michael S. "Men's Responses to Feminism at the Turn of the Century." *Gender and Society* 1:3 (1987): 261–283; Laqueur, Thomas W. *Solitary Sex: A Cultural History of Masturbation*. Cambridge, MA: Zone Books, 2003; Macomber, D., and M. Sanders. "The spermatozoa count: Its value in the diagnosis, prognosis, and treatment of sterility." *New England*

Journal of Medicine 200:19 (1929): 981–984; Martin, Emily. "The Egg and the Sperm." In Janet Price and Margrit Shildrick, eds. *Feminist Theory and the Body.* New York: Routledge, 1991, pp. 179–189; Pinto-Correia, Clara. *The Ovary of Eve: Egg and Sperm and Preformation.* Chicago: University of Chicago Press, 1997; Riddle, John M. *Contraception and Abortion from the Ancient World to the Renaissance.* Cambridge, MA: Harvard University Press, 1994; Shackelford, T. K. "Preventing, correcting, and anticipating female infidelity: three adaptive problems of sperm competition." *Evolution and Cognition* 9 (2003): 90–96; Smith, Robert L., ed. *Sperm Competition and the Evolution of Animal Mating Systems.* New York: Academic Press, 1984; Spark, Richard F. *The Infertile Male.* New York: Springer, 1988.

Lisa Jean Moore

Skin

Body Piercing

Body piercing is the practice of perforating the skin, usually followed by the insertion of jewelry. Its more specific forms such as ear piercing are among the most common types of body modification. Unlike more permanent body markings, body piercings can perish easily over time. Thus, it is difficult to determine the complete history of body piercing. However, it is clear that body piercing has been undertaken by many cultures and throughout human history. Much of the information about ancient piercing practices comes from archaeological sites, bas-reliefs, statues, other cultural objects and artwork, and a few mummified human bodies. Throughout the twentieth century, anthropologists, as well as writers for the magazine *National Geographic*, provided much information about piercing practices in traditional societies. In the late twentieth century, body-piercing enthusiasts have increased public interest in body piercing.

Body piercing includes all types, from earlobe piercings to genital piercings, as well as practices such as piercing stretching, surface piercing, and so-called play piercing, which is a temporary piercing or set of piercings lasting only a matter of hours or days. Both historically and across the globe, the most common places on the body to pierce have been the ear (especially the lobe, but also the rim of the outer ear), the nasal septum or nostrils, and areas on or around the mouth and lips. Popular contemporary piercings such as those of the navel, tongue, nipple, and female genital piercings are not known or documented in traditional societies. The tools that were used for piercing before the development of piercing guns and hollow needles included thorns, sticks, bamboo spikes, bone splinters, knives, and awls. Reasons for piercing in both traditional societies and contemporary Western culture have ranged from ritual to beautification (which may include an erotic element). It is common for both factors to be in play.

Traditional Societies

The earliest forms of piercing, according to archaeological evidence, include ear piercing (5000 BCE, China), nose piercing (5300–4000 BCE, Iraq),

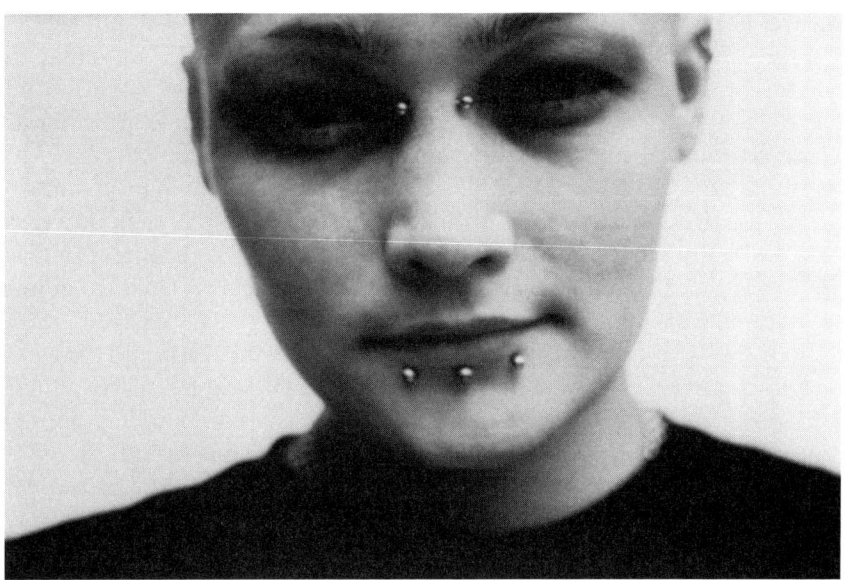
Facial piercings, including labret piercings and a bridge piercing between the eyes. © Getty Images/PhotoDisk/Tim Hall.

and lip piercing (4000 BCE, Aleutian Islands, possibly dating back to 6000 BCE). Earlobe stretching dates to roughly 1500 BCE in both ancient Egypt and Central and South America. The earliest written evidence of the male genital piercing, known as the apadravya, comes from ancient India and is found in the *Kama Sutra* by Mallanaga Vatsyayana, written in the fifth century CE. Although jewelry worn in the tongue is a recent invention, the Mayan people of the Americas temporarily pierced their tongues for ceremonial purposes.

Cultures that practiced body piercing often employed multiple types of piercings. The native people of present-day Alaska and the Aleutian Islands practiced multiple forms of facial piercing from prehistoric times until the nineteenth century. Piercing implements and stone and bone labrets, cheek plugs, and nose ornamentation have been retrieved from archaeological sites. There is evidence of lip piercing, cheek piercing, and nose piercing. In some instances, these areas were pierced multiple times. The indigenous people of Kodiak Island pierced the lower lip several times and wore small vertical labrets so that it almost appeared as though they had an additional row of teeth.

Indigenous groups in North and South America pierced as well. Multiple piercings were common among the Late Hohokam (1050–1450 CE) who lived in the southwest region of the present-day United States. They wore nose and cheek plugs made of soapstone and argillite, and made earrings of imported marine shell and stone. In central Mexico from 200 BCE to 1500 CE, elite classes wore ear flares, ear spools, nose pendants, and lip plugs. The Suya of Brazil's Mato Grasso pierce and stretch their earlobes and lips. This is done as a rite of passage among males and females, and is thought to improve both hearing and speaking. The Suya pierce their ears as adolescents and their lips as adults. Red is used for lip disks, and white for ear disks. The colors are symbolic: red is understood to be an active color, while white suggests passivity. The choice to use red for lip disks and white for ear disks helps reinforce cultural ideals in regard to hearing (and, by extension, understanding) and speaking or communicating.

Multiple kinds of body piercing are also practiced among groups in Africa such as the Dogon and the Ga'anda. The Dogon pierce their lower lips, noses, and ears. They wear twisted or weaved materials through their piercings, a practice that symbolically reinforces social ideals related to an important deity of weaving. These materials are thought to be "weaved" through the body via the piercings. The Ga'anda pierce the ears and upper and lower lips of girls as a sign that they are almost ready for marriage. Other traditional societies practice multiple piercings as well, including aboriginal cultures of Borneo and Australia and cultures indigenous to India.

Two types of body piercing that have played roles in several traditional societies are temporary piercing and flesh-hook pulling. This is most notable in the primarily Tamil Hindu Thaipusam festival, which celebrates both the birthday of Lord Murugan (also Murukan) and when his mother Parvati gave him a special *vel* (lance) that helped him triumph over evil demons. Hindus who take part in the Thaipusam festival may do so bearing a *kavadi*, which is any sort of weight or burden. This practice involves temporarily piercing the flesh with hooks that hold fruit or some other type of weight or skewers (also referred to by practitioners as *vels*)—although for some practitioners it may require only the carrying of weighty canisters. The skewers are pushed through the tongue and/or cheeks and remain in place throughout the ritual. A select number of devotees carry the larger kavadi, a framework of skewers balanced upon the chest, shoulders, and back. These skewers may pierce the flesh. Flesh-hook pulling is also practiced in this festival. This involves inserting hooks through the flesh on the upper back and pulling some type of weight—in some instances, another person will walk behind the hooked devotee and pull upon strings attached to the flesh hooks. These practices are available to both men and women; however, men most often take on the larger framework kavadi and flesh-hook pulling.

The Native American sun dance ceremony also employs the use of temporary piercing and flesh-hook pulling. This ceremony of the Plains Indians of North America (who spanned from modern-day Saskatchewan in Canada to Texas in the United States) was traditionally practiced annually in the late spring or early summer. The ceremony helped to reaffirm beliefs held in common among both settled and nomadic Plains tribes while spiritually empowering the sun dancers themselves. Some of the more elaborate and developed ceremonies required nearly a year of preparation from participants. These elaborate ceremonies included much of the community in preparation of food, space, and structures, and also the participation of spiritual mentors to the dancers.

The sun dance itself lasted several days and nights, during which participants could neither eat nor drink. Beyond the dancing until exertion, some tribes also participated in temporary piercing and flesh-hook pulling. The spiritual mentors would pierce the dancers' skin with skewers through either the upper chest or back. In some tribes, the mentor would then tie heavy objects such as buffalo skulls to the skewers. In other tribes, the piercings were used to tie the dancer to a pole. In either case, the dancer would usually dance until the piercings were torn free of their attachments. In 1883, however, the U.S. Bureau of Indian Affairs criminalized the sun dance. This prohibition was again renewed in 1904 until reversed in 1934. Despite these efforts to prohibit the ritual, the sun dance still exists and is practiced in the twenty-first century.

Another type of temporary and usually ritualistic body piercing is flesh-hook suspension, which involves putting hooks or some other type of sharp implement through the flesh and then suspending one's body in mid-air. Perhaps the most well-known example of flesh-hook suspension is the Mandan O-Kee-Pa ceremony. The most well known version of this was depicted in the movie *A Man Called Horse* (1970).

Some Hindus also practice flesh-hook suspension. These practices occur not only among the sadhus, or ascetic holy men, of India, but also among laypeople. As of the 1960s, the following practice was documented among a people of the Deccan plateau in southern India. During the full moon in April, a young, married man impersonates a patron deity of the village. During the ceremony, he is pierced with flesh hooks in his lower back. The hooks are attached to a structure, and he is swung around in large circles. After the ceremony, which is meant to be joyous, wood ashes are rubbed into his wounds.

Contemporary Western Body Piercing and the Body-Modification Movement

In the modern West, recent enthusiasm for body piercing has primarily been limited to ear piercing. However, Western mainstream society has broadened its practices from earlobe piercing primarily for women to body piercing for both men and women over a relatively quick period. This transition took place in the last quarter of the twentieth century. Some Western enthusiasts of body piercing argue that the history of the practice, along with other forms of body modification, suggests a need for bodily expression and ritual that is universal. Members of the body-modification movement who call themselves modern primitives have argued that manipulating the body is a form of "primitivism" from which all people, including those in the contemporary West, can benefit. More broadly, those who identify with the body-modification movement or subculture are interested in exploring the body as a site of manipulation, adornment, and transformation. Some consider the most recent flowering of body piercing to be a renaissance or revival. However, contemporary piercing practices have surpassed various modes of historically common piercings in terms of variety, methods, and innovation.

Contemporary piercing has built upon and expanded all manner of body piercing. There are several types of ear piercings (including earlobe stretching), nose piercings, and genital piercings, as well as various nipple, lip, cheek, eyebrow, and eyelid piercing options. Several piercings are done inside the mouth, including vertical or horizontal tongue piercing, tongue web piercing, piercing of the upper and lower lip frenulum (the web of skin connecting the inside of the lips to the gums), and uvula piercing.

One of the more experimental types of piercing is surface piercing. Generally, a surface piercing is a perforation made horizontally under the skin, with the entry and exit holes on the same plane. One of the most important elements in the success of this piercing is the surface bar, an invention most often credited to piercer Tom Brazda. This piece of jewelry is designed according to the contour of the body part being pierced. If a piercer does not use the proper method or uses the wrong jewelry, surface piercings are often rejected, which happens when the jewelry is pushed out of the skin in the process of healing. Surface piercings also include nape piercings, cleavage piercings (between the breasts), and chin piercings.

Play piercing, another outlet for piercing enthusiasts and those in the bondage/domination/sadomasochism (BDSM) subculture, often includes piercing the skin with hypodermic needles for erotic, ritualistic, or aesthetic reasons. Another expression of body piercing culture is the interest in flesh-hook suspension and pulling. Fakir Musafar (1930–), the so-called father of the modern-primitive movement, can be credited for popularizing suspension and pulling in the United States and other Western countries. In the documentary *Dances Sacred and Profane* (1987) and in photographs by Charles Gatewood, Musafar is depicted performing a flesh-hook suspension. Drawing from the history of indigenous suspension rituals like the O-Kee-Pa, contemporary suspensions are practiced for several reasons, ranging from ritual to tests of strength.

The recent history of body piercing is often traced back to innovators such as Doug Malloy, Jim Ward, and Mr. Sebastian (also known as Alan Oversby) and their activity in the gay BDSM communities in Los Angeles, London, and other cities. Ward opened his piercing shop, The Gauntlet, in Los Angeles and began publishing the magazine *Piercing Fans International Quarterly (PFIQ)* in 1977. The publication of the book *Modern Primitives* in 1986, containing interviews with and photographs of Musafar, Ward, Raelyn Gallina, and other proponents of body modification, helped to define piercing and other body modifications as meaningful, even spiritual, rather than antisocial, dangerous, or criminal. Finally, punk culture also played a part in popularizing body piercing.

There are indications that body-piercing enthusiasts were abundant in the late 1970s. In his series of articles entitled "Running the Gauntlet" (http://www.bmezine.com), Ward writes about his own experience of becoming the first person to open a professional body-piercing studio in the United States. The Gauntlet began doing business in 1975 in Ward's house in West Hollywood. In 1978, Ward realized that he was making enough money to open a studio in a commercial space. Until The Gauntlet opened its doors, professional body piercing was done as a sideline by only a handful of tattoo artists. Ward's fascination with piercing began in 1967 in the gay S/M and leather scene of New York City. A close friend pierced Ward's ears, and he found the experience intimate and transforming. This dynamic, in which close friends and sometimes lovers ritualized piercing, influenced much of the piercing scene that became popular in the 1980s.

Finding community and building trust were important components in the creation of a piercing subculture. Ward recalls that throughout his evolution as the first professional body piercer, he often found comfort in realizing that he was not alone whenever he met someone with similar interests in body piercing. To facilitate trust, community, and practical skill in the piercing subculture, Ward and other enthusiasts formed a group that met for piercing parties. Ward's magazine *PFIQ* helped to create a piercing culture by documenting a history of piercing in other cultures and presenting portraits of pierced individuals, thus giving piercing a face and a history.

Musafar, often credited with coining the term "modern primitive," is also one of the movement's most prominent figures. Musafar argues for a spiritual understanding of piercing, tattooing, and other forms of body modification. In 1992, Musafar launched the magazine *Body Play and Modern Primitives Quarterly*, introducing his philosophy to the wider public. For Musafar, body modification is an essential aspect of being human and an expression of a primal urge to modify one's body. Sometimes, Musafar argues, modification involves experiencing different states of consciousness, such as might happen in a painful flesh-hook suspension; at other times, he suggests that one might increase his or her spiritual awareness through getting a tattoo or a piercing.

In 1994, Shannon Larratt created the *Body Modification E-zine* (http://www.bmezine.com) as a personal Web site where he displayed pictures of his own body modifications. In a matter of months, he began receiving messages and pictures from other modified people. BME quickly became the most comprehensive Web site on body modification, and now includes thousands of pictures, shared stories, an encyclopedia that describes types of modifications and offers safety tips, featured articles and interviews, links to body modification in the news, and a members-only online community called IAM.

The Internet has become an important venue for the expansion of the body-modification movement. Sharing pictures and stories of body-modification experiences, readers and members generate the social interactions of an online modified community. Larratt encourages people with modifications to become more aware of themselves as a group with a shared set of interests. Larratt and other writers for BME, like lawyer Marisa Kakoulas, raise issues of individual rights and body-modification ethics, and showcase examples of body-modification artists who are innovative, politically active, or in some way personally transformed by body modification.

One of the primary concerns of body-modification leaders like Musafar, Larratt, and others is to legitimize body modification to a sometimes-skeptical public. More extreme forms of body piercing (that is, anything in excess of a female's pierced ears), for example, on skin visible to the public disrupts popular notions of "natural" beauty by lacerating the smooth and continuous surface of the skin. People who are not piercing enthusiasts therefore tend to frame body piercing as mutilative and deviant. This is often backed up by contemporary psychologists and psychiatrists, who treat body modification as self-mutilation, symptomatic of illness. Popular utilitarian notions of pleasure as good and any pain as bad also aid negative perceptions of body modification. Body-modification leaders have countered these perceptions by pointing to ancient and traditional roots of body modification and integrating contemporary body-modification practices with artistic and spiritual narratives.

Body piercing has become increasingly innovative and high-tech. Body-modification pioneers in the body-modification community have created new forms of body modification such as the 3-D art implant (an object inserted beneath the skin). These can include shapes that are visible through the skin, such as a ring, or objects that add an effect such as the appearance of horns beneath the skin of the upper forehead. Like stretched piercing, 3-D art implant also need to be inserted in stages in order to stretch out the skin for larger insertions. The transdermal implant appears as a post protruding from the skin. Three new ear piercings are called the industrial, daith, and rook piercings. The apadydoe is a new genital piercing for men that passes orbital jewelry through the corona that continues through the glans of the penis.

Some sociologists speculate that body piercing and body modification in general have become increasingly popular because of an attraction to authenticity and personal ritual in Western culture. As the grip of tradition and religion has slipped in the West, new ways of self-definition have taken their place. Body piercing and body modification are permanent and semi-permanent ways of reaching beyond mere fashion. They involve commitment in terms of overcoming pain and require more effort than changing one's clothes or hairstyle. Because of this commitment, they appear to be more serious or "authentic," which is an important ideal in boomer and post-boomer generations. In addition, as personal spirituality has replaced religious affiliation in secularized communities in the West, the ritualistic aspects of body piercing put forth by Musafar are significant. Body piercing offers a connection with what is popularly understood as an ancient practice—a connection that is perceived as adding legitimacy to the individual's choice to frame piercing as ritual. *See also* Chest: O-kee-Pa Suspension; Ears: Ear Piercing; Ears: Earlobe Stretching; and Genitals: Genital Piercing.

Further Reading: Body Modification E-zine Encyclopedia, http://wiki.bmezine.com/index.php (accessed May 2007); *Body Modification E-zine* Web Site, http://www.bmezine.com; Brain, Robert. *Decorated Body.* New York: Harper and Row, 1979; Gay, Kathlyn, and Christine Whittington. *Body Marks: Tattooing, Piercing, and Scarification.* Brookfield, CT: Twenty-First Century Books, 2002; Peregrine, Peter N., and Melvin Ember, eds. *Encyclopedia of Prehistory, Vols. 1–8.* New York: Plenum, 2001, 2002; Pitts, Victoria. *In the Flesh: The Cultural Politics of Body Modification.* New York: Palgrave Macmillan, 2003; Polhemus, Ted. *Hot Bodies, Cool Styles: New Techniques in Self-Adornment.* London: Thames & Hudson, 2004; Rubin, Arnold, ed. *Marks of Civilization.* Los Angeles: Museum of Cultural History, UCLA, 1988; Vlahos, Olivia. *Body: The Ultimate Symbol.* New York: J. B. Lippincott, 1979; Vale, V., and Andrea Juno, eds. *Modern Primitives.* San Francisco: Re/Search, 1989; Ward, Jim. "Running The Gauntlet," a series of articles. *Body Modification E-zine,* http://www.bmezine.com/news/jimward-all.html (accessed May 2007).

Jaime Wright

Branding

Branding is the process of creating a permanent scar through the use of heat against the skin, traditionally by burning the flesh with a hot metal implement. The branding of livestock such as horses or cattle to denote ownership has been documented across a wide variety of cultures since ancient times, with the earliest known documentary evidence and artifacts dating from the Egyptian Old Kingdom in the thirtieth century BCE. The branding of humans is a narrower practice, though it too has a considerable history.

Human branding has historically been undertaken primarily within a penal framework. As a corporal punishment, branding was used in a number of historical contexts to permanently mark a convicted criminal with a publicly visible sign of his or her misdemeanors, ensuring that the perpetrator would be humiliated and stigmatized. The punitive use of the practice in ancient history has been definitively documented to have been applied by the Babylonians and ancient Egyptians, and most historians agree it also was used as a means of punishment by the classical Greeks and Romans. In more modern times, the French, British, Russians, and early American colonists also embraced the use of branding of criminals and dissidents, from robbers, perjurers, and adulterers to military deserters and blasphemers. The form of the brand would usually vary depending on the specific crime, such that each in case it would be obvious what crime its bearer had committed: the English would burn the letter "T" onto the forehead of thieves, for example. With the advent of penal reform in the late eighteenth century, the popularity of corporal punishment generally began to wane, and by the 1840s, penal branding had essentially been eschewed by all developed nations.

The branding of human beings has also been historically employed by slave owners to mark their slaves. Premodern societies among whom the branding of slaves was acceptable included China, Egypt, and Rome, but it does not seem that slaves were routinely branded by their owners. In these civilizations, the branding of slaves, intended to be both a deterrent and a punishment, was restricted to those who fled captivity and were subsequently caught. Much as numerous other cultures had used unique brands to mark ownership of livestock, however, branding did come to be used simply to indicate ownership of enslaved human beings, and during the period of the European colonial slave trade from the mid-sixteenth century until its abolition in the 1830s, the use of branding was common, though not universal. Slave owners in the American colonies would often use the same branding irons that they used on their livestock to brand their slaves, and, though the practice of routine branding waned, cases of owners marking their slaves in this way have been documented as late as 1848.

Branding as a mode of religious devotion is relatively rare, although its use for religious ends has been documented among certain particularly fanatical or orgiastic sects. For example, Ptolemy Philopator, pharaoh of Egypt between 221 and 205 BCE, was branded with an ivy leaf in honor of the goddess Dionysus. The second-century CE Gnostic Carpocratians, followers of Carpocrates in Alexandria, were said to have branded their convert's earlobes as a mark of devotion to the movement.

With the rise in popularity of subcultural modes of body modification in the 1990s, the aesthetic functionality of branding achieved some prominence. In the latter decades of the twentieth century, branding was practiced in the context of sadomasochistic sexual relationships. The symbolism afforded to branding by its history had an appeal in the framework of sexually sadistic master-slave interaction, but became more widespread once activities such as piercing emerged from this community. As such, branding today is primarily an aesthetic pursuit, although its historical significance has led to its use among some university fraternities in the United States as a marker of group identity, and its perceived status as a fairly extreme form of body modification undoubtedly leads some individuals to choose to be branded as much for the experience as the resultant scar.

Contemporary branding still generally utilizes the same basic methodologies that have been used for thousands of years. A piece of metal, gripped in pliers and shaped to produce the specific wound required, is heated with a blowtorch or similar heat source until it is red-hot. The glowing implement is then pressed firmly into the skin for a few seconds, burning the flesh away and leaving a wound. Modern artistic branders tend to produce the design by branding a number of smaller strikes in order to achieve maximum control, as due to the vagaries of the healing process, the resulting brand will be noticeably larger than the implement used to inflict it.

The branding process itself is painful, but the initial strikes burn through and essentially destroy the surrounding nerve endings. The resultant healing period is relatively traumatic, however, and can take as long as six months to fully heal. The results of branding, like other forms of scarification, are unpredictable, although in general those with darker skin tend to produce more prominent scar tissue. These scars, known as keloids, will generally look fairly red and aggressive for the first month of healing, but eventually settle until they are light-colored and slightly raised, although some may disappear entirely, others may actually become indented rather than raised, and some may stay red for a considerable period of time. In order to mitigate these variations, some modification artists have experimented with other ways of burning the skin, including the use of cautery pens and electrosurgical devices known as Hyfrecators (usually used to cauterize wounds after surgery).

Further Reading: Eaton, Marquis. "Punitive Pain and Humiliation." *Journal of the American Institute of Criminal Law and Criminology* 6/6 (March 1916): 894–907; Jones, Christopher P. "Stigma: Tattooing and Branding in Greco-Roman Antiquity." *The Journal of Roman Studies* 77 (1987): 139–155; Miethe, Terance D., and Hong Lu. *Punishment: A Comparative Historical Perspective*. Cambridge: Cambridge University Press, 2005.

Matthew Lodder

Cultural History of Skin

Skin is the largest, most readily visible organ of the human body. It plays vital roles in temperature regulation, maintaining fluid balance, and keeping pathogens from entering the body. Given its central importance in several biological functions, it should come as no surprise that an entire medical subfield—dermatology—is dedicated to its healthy maintenance. Skin, however, is far from being a culturally neutral organ. Throughout history, elements of the skin's appearance (such as color and texture) have been used as important markers of race, class, and social worth. Because of the varying salience of these social markers across time and space, a variety of techniques have been developed to modify or adjust the skin's appearance.

Biology of Skin

Skin is part of the integumentary system, the system of tissues that cover and protect the body, including the skin, hair, nails, sweat glands, and their products. Aside from acting as the body's first line of defense in protecting humans from pathogens and other harmful substances, the skin plays numerous roles in maintaining homeostasis and providing the brain with sensory information about the body's surroundings. These roles include thermoregulation, excretion of wastes, maintaining fluid balance, protection from potentially harmful ultraviolet radiation, the production of vitamin D, and sensing pain, pressure, and temperature. But the skin comprises more than just skin cells. Skin is actually a complex organ containing hair follicles, sweat glands, sebaceous glands (which produce oil to keep the skin moist and supple), blood vessels, and numerous nerve endings and sensory receptors. Because of the skin's complexity and major role in numerous body functions, an entire branch of medicine—dermatology—is dedicated to its healthy maintenance and protection.

The skin is comprised of two main layers—the epidermis and the dermis. The epidermis is the outermost layer of the skin. It is comprised largely of dead skin cells, and serves as a protective barrier. These dead cells regularly slough off and are replaced, thus renewing the protective barrier. The epidermis also contains melanocytes, the cellular structures that produce the pigment melanin. There are no blood vessels or nerve endings in the epidermis; if this layer is breeched, it will not hurt or bleed. The dermis is the innermost layer of skin, and contains blood vessels, nerve endings, sweat glands, lymph tissue, hair follicles, and sebaceous glands. If this layer of skin is breeched, one will experience a painful sensation, and bleeding is possible.

While the basic structures of the skin are universal in healthy people, many qualities of the skin vary between individuals. Marked differences can be observed in skin color, opacity, thickness, and firmness across varying climates, geographic locations, and age groups. Some of these differences and the cultural meanings attached to them are discussed below.

The Aging Process

Aging skin differs from young skin in numerous ways. Over time, the skin becomes thinner and more delicate. As metabolism slows with age, the rate at which the skin renews itself decreases. There is less blood flow in aging skin, and the production of sebum (oil from the sebaceous glands) slows markedly. Collagen and elastin (naturally occurring proteins primarily responsible for skin's firmness and tautness) are produced in smaller quantities, and existing stores begin to break down. These changes can cause skin to become blotchy, sag, or wrinkle.

Given the premium placed on a youthful appearance in many parts of the world, various dermatological products and procedures have been developed to counteract aging's effects on skin. In fact, some anti-aging skin treatments date back to historic times. Ancient Egyptians are noted for their use of cosmetics, and numerous papyri are filled with recipes for various styles of makeup and cosmetic treatments. One such recipe consists of a mixture of waxes, incense, olive oil, fresh milk, and cypress, which was applied to the face over the course of six days. During biblical times in the Middle East, oils and extracts from olives, sesame seeds, and a variety of fruits and nuts were used to rejuvenate and moisturize the skin, as well as protect it from sun damage. Ancient Greeks were known to use a poultice of bread and milk to reduce the signs of aging skin, and Romans are credited with inventing cold cream, a compound used to smooth skin and remove makeup, around the second century CE.

Since these early innovations, humans have developed much more high-tech approaches to maintaining firm, youthful, glowing skin. Among the most technologically and medically advanced anti-aging dermatological measures in the West is the facelift, a cosmetic surgical procedure designed to stretch the skin more tautly across the face and neck. Such surgeries reduce the appearance of sagging or wrinkles (a similar procedure can be performed on various other body parts to give the skin a smoother, firmer, and younger appearance overall). Aside from cosmetic surgery, one of the best known and most widely used anti-aging skin treatments in the West is Botox. Designed to be administered via injection by a licensed dermatologist, Botox is a serum derived from the bacterium that causes botulism. Generally injected into facial tissue, Botox temporarily paralyzes superficial muscle layers. The visible effects this produces include decreased appearance of deep wrinkles and the softening of lines in the skin covering the affected muscles. While Botox injections do reduce some of the visible signs of aging skin, it has also been noted that the capacity to produce some facial expressions may be lost due to the paralysis of certain facial muscles.

Another, less permanent anti-aging skin treatment is Thermalift. Also designed to be administered by a licensed dermatologist, Thermalift is a nonsurgical procedure that helps to temporarily firm skin by using radio waves. Purportedly, the radio waves penetrate to the deepest levels of skin and stimulate natural collagen production. However, this procedure must be repeated at regular intervals to see consistent results, and as such can be quite costly.

Rather than opting for treatments to stimulate their own collagen production, those seeking to reduce signs of aging skin may opt to receive collagen injections. Although collagen is a protein naturally produced in human skin, collagen used for injections is generally derived from cattle. This harvested collagen is processed and injected into sunken areas to firm and fill out wrinkles or sagging skin. Collagen is often also used to give the appearance of plumpness or fullness, such as when injected into the lips.

Wrinkling and sagging are not the only issues that concern those who wish to mask the appearance of aged skin. Many people also wish to address the appearance of dullness or discoloration that can result from the decreased blood flow, lower sebum production, and slower skin renewal as experienced with increased age. One popular approach to addressing this problem is microdermabrasion. As the name would suggest, microdermabrasion uses either jets of a chemical solution, specially formulated powders, or an abrasive surface to exfoliate, sand, or buff away the uppermost epidermal layers. This makes brighter, younger-looking layers of skin more visible. Microdermabrasion is not to be confused with dermabrasion, which uses a much harsher abrasive and is generally done to remove scars.

In addition to major cosmetic procedures that must be performed by a licensed dermatologist, many products and procedures have been developed to reduce the signs of aging skin. Many people opt to use makeup to cover so-called "age spots" (areas of hyperpigmentation that can be especially visible in lighter-skinned individuals as they age), mask the appearance of fine lines and wrinkles, or give their skin a brighter, younger tone. Countless products have been marketed specifically to those looking for a more youthful appearance. A number of over-the-counter products have been designed to slough off superficial layers of skin, increase firmness, or reduce wrinkles. Some of the most popular at-home anti-aging treatments contain alpha-hydroxy acids, a group of chemical compounds that cause the outermost layers of skin to slough off, much as would be done with microdermabrasion. When used in higher concentrations, alpha-hydroxy acids must be applied and monitored by a dermatologist in a procedure known as the "chemical peel." Many other products contain antioxidants and/or sunscreens to either repair or prevent skin damage from sun exposure.

Skin Color

The skin's features can vary greatly across social groups. One of the skin's most readily visible attributes is its color. Skin is given its color largely by melanin, a type of pigment naturally produced by the melanocytes in healthy human skin. The amount of melanin in one's skin is determined largely by two main factors—one's genetic makeup and environment. Human skin generally contains two subtypes of melanin—the brownish-black eumelanin, and the reddish-yellow pheomelanin. Variations in their respective concentrations are responsible for differences in skin color, shade, and tone. While variations in the ratio of eumelanin and pheomelanin result in varying shades or tones, it is a general rule that higher concentrations of melanin result in a darker skin color, while lower concentrations yield a pale skin color. As such, one can imagine two individuals with the same skin tone (that is, the same ratio of eumelanin to pheomelanin) but with markedly different skin colors (that is, one being much darker than the other). The complete absence of melanin is the result of a genetic condition known as albinism.

Genes

Several genes affect skin color. While some genes have indirect effects on skin color—such as skin's opacity or brightness—others directly affect skin color. In particular, several genes influence the number and type of melanocytes one will have in one's skin. In general, the greater the number of melanocytes, the greater one's capacity to produce melanin. The best-studied of such genes is melanocortin-1 (also known as the Mc1r gene), which plays a major role in both hair and skin pigmentation. While the genetic component of skin color can be quite complex, scientists have suggested that as little as 6 percent of human skin-color variation is the result of genetic factors with the remaining variance being the result of environmental factors.

Environment

The melanin present in human skin serves more than an aesthetic purpose. One of melanin's chief protective functions is to filter out potentially harmful ultraviolet (UV) radiation. When human skin is exposed to direct sunlight, it is bombarded with UV rays. In sufficient doses, UV radiation can lead to sunburn, skin damage, and, with prolonged exposure, even skin cancer. Fortunately, the melanocytes in human skin are stimulated to increase melanin production in the presence of ultraviolet light, thus increasing the skin's ability to protect itself from damage. This is the biological foundation for the process known as sun tanning.

Anthropologists have posited that environmental factors are largely responsible for the patterns in skin-color distribution that one can observe in the people of the world. Specifically, climate and UV exposure vary by latitude, with increased exposure near the equator, and gradually decreasing exposure approaching the poles. Given this, anthropologists have posited several theories as to why different skin colors may have adaptively evolved in varying parts of the globe. One such hypothesis is the "vitamin D hypothesis." Vitamin D is a biochemical building-block that plays central roles in calcium metabolism, bone formation, and immune function. While some vitamin D is consumed from dietary sources, it is chiefly produced when the skin is irradiated by ultraviolet light. Since melanin acts as a filter for ultraviolet radiation, many researchers have argued that darker-skinned individuals would have been at an evolutionary disadvantage at higher latitudes where sun exposure and UV intensity are relatively low. Lighter-skinned individuals would have been better suited to such an environment, given that their skin produces little melanin, allowing for optimization of the available UV rays for vitamin D production. In this same vein, it has been suggested that lighter skin would place an individual at an evolutionary disadvantage at latitudes where sun exposure and ultraviolet intensity are high. One of the possible results of overexposure to UV radiation is sunburn. Aside from being red, hot to the touch, and quite painful, sunburned skin suffers from a decreased ability to sweat, and its role in thermoregulation is compromised. In tropical climates, such a condition could quickly prove fatal. The amount of melanin present in darker skin tends to protect one from sunburn, making darker skin a favorable characteristic in environments with intense UV exposure.

Most anthropologists believe that original humans were most likely dark skinned. By and large, the scientific community agrees that human life originated in Africa. The historical environmental conditions present in Africa, the protective advantage of high concentrations of melanin in tropical climes, and genetic studies of early human remains suggest that all humans were initially dark skinned. Lighter skin evolved as humans migrated to other parts of the globe and environmental pressures or needs changed.

Race and Skin Color

While the biology behind skin color may seem rather cut and dry, human beings have made a variety of distinctions between groups based on skin color. Perhaps the best known of such man-made distinctions is the categorization scheme called "race."

While race and skin color are synonymous in the minds of many, the two were not always linked. During the Enlightenment, beginning in the seventeenth century, Europeans were increasingly obsessed with quantifying, classifying, and explaining the world around them. While certainly true in terms of the natural sciences (such as biology or chemistry), this bore itself out in terms of human interactions as well. During this era, there was an increased emphasis on investigating and describing the whole of the earth. Countless explorers took up the task of sailing around the world, mapping the terrain, and describing in excruciating detail the flora, fauna, places, and people they encountered. These accounts frequently included very detailed descriptions of the physical characteristics of people, ranging from temperament to the color of their skin. Initially, no more salient than any other physically observable characteristic, skin color was first used as a mere descriptor. Many tribal groups throughout Asia and various other parts of the Northern Hemisphere were initially described as "white-skinned."

By the 1770s, skin color ceased to be thought of as a descriptive term, and began to be cast as immutable and inherently associated with a variety of personality or cultural attributes. The traits of lighter-skinned groups tended to be seen favorably, whereas the attributes of darker-skinned peoples were seen as savage, backward, uncivilized, and unrefined. As this thinking emerged, the term "race" began to take on the shape of the concept as it was used for the majority of the nineteenth and twentieth centuries.

For both anthropologists and dermatologists, the scientific assessment and classification of skin color has been an important endeavor throughout the history of the respective disciplines. While dermatologists have been interested in this venture for reasons pertaining to the maintenance of healthy skin, anthropologists have historically been interested in the scientific measure of skin colors for the purposes of separating human beings into different categories or groups. Anthropologists once used the broad "racial" or skin-color categories of black, white, yellow, red, and brown— with each of these color categories being paired with various other physical characteristics. This classification system dominated the discipline until the late nineteenth/early twentieth century, when the von Luschan scale gained favor. Created by German doctor and anthropologist Felix von Luschan (1854–1924), the von Luschan scale consisted of a series of opaque colored tiles in a variety of hues. The tiles were held up and matched to "unexposed" areas of the subject's skin—such as the upper inner arm—where the skin was likely to be untanned. Highly popular until the 1950s, this method for quantifying skin color was largely replaced by spectrophotometry, an electronic measurement of skin's spectral reflectance.

In much the same vein, explorers and anthropologists often endeavored to delineate which skin colors/types could be found in varying parts of the globe. A product of this was the skin-color map, the most famous of which was created by Italian geographer Renato Biasutti (1878–1965). Biasutti's map, based on the von Luschan scale, placed darker-skinned people near the equator, with skin colors becoming incrementally lighter approaching the poles.

The common thread uniting these detailed travel diaries, skin-color maps, and the skin-color scales is that they were born of the drive to classify human beings based on skin color. Each of these approaches, however, was anecdotal and imprecise at best. Perhaps the most "scientific" of racial/skin-color classification schema is that developed by Carolus Linnaeus (1707–1778), an eighteenth-century Swedish physician and biologist. He is credited with developing and popularizing the major classification/taxonomic system still used by biologists today. This system allowed biologists to separate all life forms into increasingly specific categories based on their shared characteristics. While all known life forms each belong to a kingdom, phylum, class, order, family, genus, and species, most life forms are referred to simply by their genus and species (binomial nomenclature). Human beings are no exception to this rule, *Homo sapiens* being the genus and species of mankind. In addition to the general classification system, Linnaeus suggested that people be grouped by "race" as well, with races further specifying types within the species. Linnaeus divided *Homo sapiens* into four racial groups—*Africanus, Americanus, Asiaticus,* and *Europeanus.* His classifications were initially based on place of origin, but as his work evolved, these classifications were based increasingly on physical attributes—skin color being chief among them. The Linnaean system of classification lent racial classification systems an air of biological or scientific credibility, allowing for numerous claims about the nature of races and the separation of humankind. It was during this era that skin color became concretely married to personality or cultural attributes and conflated with the idea of "race" in the minds of the masses.

The thinking that emerged during the Enlightenment readily lent itself to the desire to quantify human attributes and make classifications based upon these observations. However, the period leading up to and continuing well beyond the Enlightenment wasn't filled with exploration and quantification alone. The boom in exploration and quantification during this time was frequently coupled with conquest, pillage, and enslavement. While slavery and colonization have existed in various cultures throughout recorded human history, chattel slavery and colonialism took on a distinctly racialized rhetoric during the Enlightenment. The "discovery" of foreign lands yielded new knowledge about the world, its climes, and its resources. It also exposed European explorers to what they saw as "unused" land and "untapped" natural resources. These explorers thought it was not only their right but their duty to teach natives in these areas how to properly use their resources and conduct their affairs in a more Eurocentric fashion. Moreover, Europeans thought it was just to take a significant portion of the land's resources in the name of the crown. This, in essence, is colonialism.

Frequently, the manual labor of those native to these "new lands" was counted among the untapped natural resources. Given the racialized rhetoric that began to emerge during this period—including likening certain human "subgroups" to animals—the wholesale misappropriation of the labor of indigenous persons was scientifically justified in racist terms. As such, many groups of people were openly captured, bought, sold, or traded and used to perform backbreaking tasks, both in their homelands and abroad. Since these individuals were considered inherently subhuman (as evidenced by their dark skins), compensation was not seen as necessary. The provision of basic survival needs—similar to those provided to livestock and beasts of burden—was thought to be sufficient. This is the essence of chattel slavery.

During this era, both colonization and slavery were attended by the denigration of all things indigenous and the valorization of all things European—including customs, language, clothes, religion, and physical appearance. Given that human taxa are "scientifically" ordered by physical attributes (as with Linnaeus' system of human taxonomy), people with physical or cultural attributes that resembled those of European colonizers were seen as being "more civilized," and thus more desirable. In colonial contexts, lighter-skinned natives were generally given positions with better pay, increased privileges, and a higher degree of autonomy. This process was paralleled in the slave trade. Lighter-skinned slaves were frequently given coveted in-house jobs, which were less severe in workload compared to the backbreaking field labor generally performed by their darker-skinned counterparts. Lighter-skinned slaves were also occasionally afforded training in the skilled trades (smithing, carpentry, etc.), as they were assumed to be more docile and intelligent. Because of the social conflation of lighter skin, intelligence, civility, and docility, lighter-skinned Africans and African Americans almost universally commanded a higher selling price on the slave market.

In many cases, lighter skin also implied a direct or familial connection to a European colonizer or slave owner. In colonial or American antebellum contexts, many lighter-skinned people of color were the illegitimate children of white male colonizers or slave owners and female concubines, or white male master/colored female slave unions. This history is painful for many communities of color because not all of these unions were consensual, but rather instances of sexual victimization of slave women. In some cases, the fathers would attempt to make some sort of provisions for their offspring, either via providing schooling, training in a skilled trade, or direct provision of money and other goods. In some instances, the offspring of a multiracial union would be light-skinned enough to "pass" for white, allowing them to take advantage of social amenities usually off-limits to people of color. Lighter skin among people of color eventually came to be viewed as an asset, because it was associated with the opportunity for greater social mobility and freedom in the opinion of the colonized and enslaved, and increased intelligence, civility, and even beauty in the eyes of colonizers and slave owners.

The ways in which skin color was viewed during the colonial and slave periods has had lasting effects. Skin-color stratification—the differentiation between persons based on lightness or darkness of skin tone—and colorism—the systematic privileging of lighter-skinned individuals within a community of color—are widespread across the globe. In several current studies of the correlates of economic and educational attainment, the highest levels of attainment were linked to lighter skin for both African American and Latinos. This pattern has been observed in communities of color throughout the Americas, Europe, Asia, and parts of Africa. Other U.S.-based studies have shown that both African Americans and whites associate various stereotypes with African Americans of varying skin tones, with lighter-skinned individuals frequently being stereotyped as more attractive, intelligent, kind, and honest, and darker-skinned individuals being seen as troublesome, untrustworthy, and prone to violence. These stereotypes have been shown to affect not only intimate but public and professional interpersonal interactions to varying degrees.

Numerous studies suggest that skin-color stratification is most apparent among women of color. Throughout Europe and the West, light skin has long been considered a desirable feminine attribute. Not surprisingly, then, skin color has been shown to be among the most salient factors in body-image evaluations for women of color. The same does not appear to hold for whites.

Class and Skin Color

While skin color has been used for centuries to divide human beings into distinct "races," skin color has also been used to divide people into class groupings—often even within racial groups. Particularly in preindustrial societies, pale or "fair" skin was a mark of nobility and high class standing. In agrarian or feudal societies, the majority of manual labor was done outdoors. Those of the lower-class strata, who were forced to work outdoors, were subject to hours upon hours of braving the sun and other harsh elements. Their skins were frequently tanned, dry, sunburned, or cracked. In fact, it is this propensity to sunburn after hours of doing manual labor in open fields that spawned the use of the term "redneck" when referring to lower-class white field hands in the American South. Those of the upper-class strata, on the other hand, were generally afforded the privilege of performing work or leisure activities indoors, where their skin was protected from the elements. Untanned, unburned, soft, and supple skin became associated with a certain degree of class privilege and leisure and was viewed as attractive and desirable. Due in large part to this association, parasols—small umbrellas carried as protection from the sun—were popular accessories for upper-class ladies who dared venture into the sun. Similarly, the use of face powders and other cosmetics to make the skin appear pale were quite common in many societies, including among European nobility and the geisha of Japan.

While sun-darkened skin was once a sign of low class standing in many societies, tanned skin has taken on new meaning in contemporary contexts. Since the Industrial Revolution, the majority of work—manual labor, skilled trades, and white-collar jobs—has been done indoors. Where laborers were once subject to seemingly endless hours of toiling in the sun, the majority of workers are now confined to factory floors or office cubicles, where sun exposure is scarce. Most workers are no longer visibly set apart by the condition or color of their skin. It is now the upper classes who may be set apart by the presence of tanned skin. Tanned skin is now identified with leisurely lifestyles filled with vacationing in sunny, tropical locales, lounging poolside, or idling on the deck of a cruise ship, sailboat, or yacht. In many contemporary contexts, a tan has become a marker of wealth, luxury, and the leisure class.

Bleaching, Tanning, and Modern Aesthetics

Given the many ways in which skin tone and color have been differently valued across time, space, and context, it is no small wonder that humans have developed so many techniques to alter the skin's appearance. These include bleaching and tanning.

Skin Bleaching

Given the fact that darker skin is still devalued in many contexts, it should come as no surprise that skin-bleaching techniques are widespread in places where darker skin is the most devalued. Skin lightening (also known as skin whitening, skin bleaching, or skin depigmentation) is the process by which the skin is chemically made to appear lighter in color, either by breaking down existing melanin stores, or by blocking further production of melanin. While many legal commercial skin lighteners are currently available in the United States and the world over, many people resort to harsh chemical mixtures or compounds of unknown composition to bleach their skin. While these products do produce a lightening effect, various—often severe—skin and systemic problems may result from their use.

Most commercially available skin-lightening products contain one of three major components (or some combination thereof): hydroquinone, corticosteroids, or mercury. Hydroquinone is a chemical compound that inhibits melanin production. Corticosteroids are a class of anti-inflammatory drugs that, in sufficient concentration, have been shown to lighten skin. However, with prolonged use, corticosteroids can thin the skin and affect its ability to repair itself. Mercury is a heavy metal and is highly toxic. Its use in body products is illegal in most countries, but it frequently appears as an ingredient in skin-lightening products sold on the black market. Because the action of creams and topical solutions is not permanent, most bleachers must continually apply bleaching agents to their bodies over a period of years to achieve some sustained effect.

In addition to the many commercially available prescription and over-the-counter skin-lightening products, homemade and black market skin-lightening products and methods abound. These can range from the use of household or industrial bleaches, detergents, and a variety of other chemicals. Many products available on the black market are assumed to be quite dangerous, as their composition is generally unknown.

Tanning

With the increase in the esteem afforded suntanned skin, there has been a marked increase in methods and mechanisms for obtaining a tanned look. Among these are tanning salons, tanning lotions, and self-tanners. Tanning salons generally allow patrons to trigger their natural tanning response with the use of ultraviolet tanning lamps. These lamps bombard the skin with ultraviolet radiation under controlled circumstances, allowing the patron to control the extent of their tan as they see fit, regardless of weather or natural climate. Tanning lotions are specially formulated creams designed to magnify the effect that ultraviolet light has on the skin. These lotions can be used either in tanning salons or when tanning in natural sunlight. Self-tanners differ from earlier methods in that they temporarily, artificially give skin a darker color. This can be achieved by the inclusion of pigments or chemicals that cause the skin to otherwise darken. Many people choose not to use this method because self-tanners can purportedly yield an unnatural orange color. *See also* Skin: Skin Lightening.

Further Reading: Altalbe, Madeline. "Ethnicity and Body Image: Quantitative and Qualitative Analysis." *International Journal of Eating Disorders* 23 (1998): 153–159; Aoki, Kenichi. "Sexual Selection as a Cause of Human Skin Color Variation: Darwin's Hypothesis Revisited." *Annals of Human Biology* 29 (2002): 589–608; Hunter, Margaret. "'If You're Light, You're Alright': Light Skin Color as Social Capital for Women of Color." *Gender and Society* 61 (2002): 175–193; Jablonski, Nina. "The Evolution of Human Skin and Skin Color." *Annual Review of Anthropology* 33 (2004): 585–623; Maddox, Keith, and Stephanie Gray. "Cognitive Representations of Black Americans: Reexploring the Role of Skin Tone." *Personality and Social Psychology Bulletin* 28 (2002): 250–259; Skin Care e-Learning and Resource Center. Anti-Aging Skin Care Resource Center for Facial Rejuvenation, http://www.skincareresourcecenter.com (accessed July 2007); Stüttgen, Günter. "Historical Observations." *Clinics in Dermatology* 14 (1996): 135–142; Wheeler, Roxanne. "The Complexion of Desire: Racial Ideology and Mid-Eighteenth-Century British Novels." *Eighteenth-Century Studies* 32 (1999): 309–332.

Alena J. Singleton

Cutting

While cutting the skin in scarification rituals is used for bodily adornment in many tribal customs, the practice of repetitively cutting the skin can be a sign of mental illness in Western cultures. Cutting, a behavior that presents in about 1 to 4 percent of the U.S. population, falls under the category of self-injurious behaviors (SIB), which are defined by the American Psychiatric Association as attempts to alter a mood state by inflicting physical harm serious enough to cause tissue damage to one's body. Self-injury is highly associated with borderline personality disorder (BPD) as defined in the DSM-IV, the fourth edition of the *Diagnostic and Statistical Manual for Mental*

Disorders, the primary diagnostic reference text for psychiatrists. While it is not recognized as its own category in the DSM-IV-TR, researchers in the field are advocating for its inclusion in the next edition of the DSM (which is not due for completion until 2010 or thereafter). Currently, self-injury is considered, as far as diagnostic standards dictate, a symptom of one or more psychiatric illnesses: depression, personality disorders (especially borderline personality disorder); bipolar disorder (manic depression); mood disorders (especially major depression and anxiety disorders); obsessive-compulsive disorder; and psychoses such as schizophrenia.

Cutting, the style, method, and tools for which vary from person to person, presents in significant comorbidity with alcohol and substance abuse, and nearly one-half to two-thirds of cutters may also have eating disorders. One explanation for cutting, as in other self-abusive behaviors, is that cutters are unable to express anger toward others, and so they turn it on themselves. In many ways, cutting can be understood as an addiction to exchanging emotional pain for physical pain. The most common reasons given for self-cutting are depression, and half of female cutters report self-punishment as their intent. Other reasons include the attempt to relieve intolerable tension, provide distraction from painful feelings, decrease dissociative symptoms, and block upsetting memories. Cutting is distinct from suicidal forms of self-harm, and is sometimes called "delicate self-harm syndrome."

Sometimes, cutters have experienced trauma in childhood. However, the correlations between childhood abuse and self-harming behaviors are not yet fully understood. It is helpful to think of self-harm (without suicidal intent) as a coping mechanism; it is more similar to an eating disorder or drug or alcohol abuse than as reflective of suicidal ideation. This distinction has great implications for treatment. Addressing suicidal ideation is very different than addressing a behavioral coping mechanism, which cognitive behavioral psychotherapies have proven effective in treating. To understand what a cutter is trying to accomplish through the enactment of her or his behaviors is to be more effective in trying to replace them with healthier, safer, more productive coping mechanisms.

Although people who self-harm are at a high risk for suicide, researchers and clinicians in the field emphasize the difference between a suicide attempt and an incidence or pattern of self-harm. The most important distinction is that of intent. In a recent study that interviewed survivors of near-fatal self-harm, only two-thirds reported antecedent suicidal thoughts. This framework suggests that symbolically speaking, cutting is viewed as a method to communicate what cannot be spoken. The skin is the projected canvas, an encasement of sorts, where aspects of the psyche reside. Cutting is poorly understood, under-researched, and arouses aversive feelings in both health professionals and the general public. As such, it tends to be conflated with suicide and burdened by inaccurate assumptions about intent. At present, little is known regarding etiology, course, diagnosis, assessment, and appropriate treatment interventions for cutting. The data available focuses on self-injury behaviors as a whole. In the meantime, there has been considerable effort to develop more accurate ways of thinking about and understanding this category of behaviors. "Parasuicide," the term used by the World Health Organization, includes all suicide methods but recognizes the range and variation of intent. Rather than implying the nature of intent, or avoiding intent altogether, this term simply avoids *ascribing* intent and leaves room for what is not understood and what cannot be assumed about the person engaging in these behaviors.

Some scholars have suggested that cutting is hard to understand because, in the medical setting, the physical symptoms take such precedence over the psychological pain that the significance of this problematic psychology is almost eclipsed, lost in the rhetoric of diagnostic parameters. Both figuratively and literally, the cutter is inscribing his or her pain onto his or her body. In this way, the body becomes the medium through which to communicate pain, however silently, and the canvas on which to inscribe this pain; physical boundaries, specifically the skin, become a point of transgression in an effort to express, alleviate, and even blunt feelings that have become excruciating, persistent, or ubiquitous within the context of the person's life.

As a coping mechanism, cutting reenacts and centers around an attention to and an articulation of the boundaries of the body. In many cases, cutters have expressed a desire to prove their humanity to themselves and to assert ownership over their bodies. Postmodern interpretations attempt to reframe self-harm as a coping mechanism that accomplishes something productive for its "user" in the short term, albeit dangerous and self-defeating in the long term. This shift in perspective may enable medical and psychological practitioners to recognize a sense of strength and will in their patients.

As studies show, self-harmers appear to have a different demographic profile than do those who commit suicide. Generally speaking, self-harm is more common in women than in men, and the most common age for onset in the United States is sixteen years. Risks of self-harm and suicide may be higher in people or communities with precarious employment situations, though uncertainty remains.

Further Reading: Anzieu, Diedier. *The Skin Ego.* New Haven, CT: Yale University Press, 1989; Levenkron, Steven. *Cutting: Understanding and Overcoming Self-Mutilation.* New York: W. W. Norton, 1998; McAllister, M. M. "In Harm's Way: A Postmodern Narrative Inquiry." *Journal of Psychiatric and Mental Health Nursing* 8 (2001): 391–397; McAndrew, Sue, and Tony Warne. "Cutting Across Boundaries: A Case Study Using Feminist Praxis to Understand the Meanings of Self-Harm." *International Journal of Mental Health Nursing* 14 (2005): 172–180; Skegg, Karen. "Self-Harm." *The Lancet* 366 (2005): 1,471–1,483; Focus Adolescent Services, "What is Self Injury?" http://www.focusas.com/Selfinjury.html; Naked Medicine, http://www.nakedmedicine.com; S.A.F.E. Alternatives, http://www.selfinjury.com.

Alana Welch

Scarification

Scarification is the deliberate cutting, branding, removal, abrasion, or use of chemicals or a cauterization tool on the skin with the intension of creating a scar. Scarification is often referred to in anthropological works as "cicatrization," which essentially refers to the forming of connective tissue over a wound, creating scar tissue.

Beauty and ritual are common themes in the practice of scarification. In traditional societies and contemporary subcultures, scarification designs can range from simple slashes or dots to intricate, technically difficult patterns. Different methods and tools can be used to achieve scars of various lengths, widths, and shapes. Some scars are raised while others are indented; the effect depends not only on the cut or brand, but also on the treatment of the wound.

History and Culture

The history of scarification dates to prehistoric times. There is an arguable case for scarification—possibly in conjunction with rites—as far back as 5000 BCE among the aborigines of mainland Australia and Tasmania. However, it is difficult, perhaps impossible, to pinpoint when scarification entered the human repertoire of body modification. Much of the discussion about the practice is based on rock art and figurines in connection with typical practices of indigenous peoples.

Modern scarification on a young Mexican woman, 2006. (AP Photo/Eduardo Verdugo)

As skin tissue often deteriorates, it is difficult to find hard evidence of scarification. The case of Ötzi the Iceman may be one of the few exceptions. Ötzi, named after the Ötztal Alps, was found in the early 1990s. His body was preserved in ice and is estimated to date to 3300 BCE. There are marks along his spine and ankles. These are often referred to as tattoos, but they may be brandings or herbal or ash rubbings. Upon first viewing the body of Ötzi, some considered the marks on his back to be brands. Others have speculated that the marks might have been made for therapeutic reasons. If this was the case, they may be cuts wherein herbs or ash were rubbed, thus causing coloration.

Scarification practices stretch not only back in time, but also around the world. In South America, the Andean Regional Development tradition dating back to 250 BCE practiced facial scarification, which was possibly used to express one's ethnic origin or socioeconomic status. In Mexico, the Veracruz practiced scarification in the first millennium CE.

By far the most documented practices of elaborate systems of scarification come from anthropological reports and studies of African traditional societies from the eighteenth to the twentieth centuries. Typically, scarification is more effective than tattooing upon skin with darker pigmentation. This has to do with a tendency of darker skin to form keloids or hypertrophied scars more easily than skin with little pigmentation.

Scarification is often used in traditional cultures to signify status, social structure, genealogy, and personality; it may also mark one as civilized or human. Some peoples have altered their bodies in order to differentiate themselves from animals. For example, the Bathonga of Africa scarify their bodies in order to differentiate themselves from fish.

Many African practitioners have described scarification as beauty enhancing, and it is often widely appreciated for its erogenous quality. As mentioned above, scarification may be used therapeutically as well. For example, until the 1960s, the Bangwa of Cameroon would cut the shape of a star with four points over the liver to prevent a man from getting hepatitis. Scarification is often multivalent; scarification marks and practices have multiple meanings and values.

Rites of Passage

In traditional societies, scarification is usually in relation to a rite of passage. In general, scarification is performed from puberty until marriage or the birth of the first child. It is common among males and females. However, the most elaborate and lengthy scarification rituals appear to be more common among females, especially in various cultures in Africa.

Among the Nuba, the scarification of girls is more elaborate than that of boys and takes place in three stages during their physiological development: breast development, first menstruation, and weaning of the first child. After these three stages, a woman has scarification patterns that cover her entire torso, back, neck, back of the arms, buttocks, and back of the legs to the knee. The courage to face difficulty is an important part of these rites. It is common for scarification to be used to test one's endurance and fitness for one's new position in life.

This cutting is done away from the village because a woman's blood is thought to be evil and polluting, an idea that is prevalent among cultures that practice scarification. An older woman of the tribe performs the cutting. The girls who undergo scarification stay away from the village until their cuts have fully healed and become scars.

The instruments used to make these series of scars are a hooked thorn used to lift the skin and a small blade used to slice the lifted skin. The higher the skin is pulled up when cut, the higher the scar and the longer it will remain visible. This method produces raised scars or keloid scars, which are thought to be the most attractive.

The Ga'anda, a Chadic-speaking people living in northeast Nigeria, also have an elaborate regimen of scarification for females. Scarification of females in Ga'andan culture starts when they are girls and continues into womanhood; the primary function is to prepare girls for marriage. The scarring is done on both the body and the face. As the scarring becomes more elaborate, the suitor's payments to the wife's family increase.

The first set of scars is made on the stomach. Two chevrons are made above the navel, which draws attention to the womb; this also signals the start of bridewealth payments from the groom. The successive stages of scars cover the forehead, forearms, torso, thighs, buttocks, and the nape of the neck. The cuts are rubbed with powdered red hematite, thought to encourage beautifully raised scars. These rituals prepare the girl for marriage both physically and emotionally; in addition, they encourage sociocultural integration.

The Gbinna and Yungar of Niger-Congo, both Ga'anda neighbors, also practice these scarification rituals. Such scarification rites for females were also practiced by the Middle Sepik of Papua New Guinea, who scarred young girls upon their breasts, bellies, and upper arms. The Imoaziri of Manam Island also scarified girls during their first menstruation, when they would receive cicatrices on the chest and back.

Males undergo similar rites of passage. The Kagoro of Papua New Guinea practice scarification as a rite of passage for young men. This helps not only to mark, but also to produce young adult males by symbolically purging any female blood from the body. Two circles are cut around the nipples of initiates, after which other circles are cut on the right and left sides of the body in a symmetrical design including the arms, chest, and abdomen. Serpent-like patterns are then cut from the shoulders down to the thighs. The cuts are rubbed with oil and mud to help create raised scars. The ultimate effect is a reptilian appearance. Large reptilian monsters represent the masculine principle among the Kagoro.

The Ache males of South America undergo the process of scarification to show they are ready to marry. They are given deep cuts down the length of their backs. A mixture of ash and honey is rubbed into the cuts to help make the scars more visible. This practice is understood as "splitting" the skin, which is related to the Ache practice of splitting the earth with axes, thought to aid in protecting order from chaos.

Aesthetics

Ritual is not the only reason for the practice of scarification. In some traditional societies, the primary purpose is aesthetic. For the Tiv of the Benue Valley of Nigeria, it is common for both men and women to undergo the process of scarification for purely aesthetic reasons. Men tend to decorate their chests and arms, while women prefer scars on their backs, legs, and especially their calves. For the Tiv, scarification designs vary with generations in much the same way that fashions change in the West. These changes in designs of one's scars can indicate one's generation among the Tiv.

Geometrical designs along with representations of fauna are common for both sexes. These designs are also common on other works of art, objects, and structures in Tiv culture. Rather than being scarified at certain predetermined stages of life, the Tiv choose to be scarified at personally significant times.

The Tiv use several methods for producing scars. They use materials such as thorns to pull the skin and blades to cut it in order to produce keloid bumps. In addition to using a particular blade for flat scars, they use nails and razors for deep scars, sometimes cutting over existing scars with nails to make them more prominent. The nail method for deep scars is used often on the face in order to accentuate facial features that accord with standards of beauty, such as prominent cheeks and long noses. Nails and razors can be used to produce several tiny lines of various lengths to create bold outlines and other features. Substances such as ashes, charcoal, camwood, or indigo are often rubbed into the wounds to make the scars more prominent.

For some traditional societies like the Yoruba in Benin and Nigeria, aesthetics are bound up with deeper philosophical concepts; scarification conveys the concept of the generating force in the world, *ase*.

Yoruba scarification is called *kolo*. This specific type of scarification is composed of short, shallow, closely spaced cuts into which pigment or charcoal is rubbed. These cuts are made with a special Y-shaped blade. This blade needs to be very sharp, and scarification artists work in a rapid rhythm. Those who perform the scarification are highly respected. These artists refer to their work as playful. However, this playfulness requires great skill and technique, the complexity of which can be deduced from the precise terms the Yoruba have for specific cutting techniques and types of scars. These variations include subtle effects such as differences in the depth, width, visibility, height, and frequency of lines, as well as in pattern, texture, color, and overall composition.

The appeal of scarification for the Yoruba is not just visual, but also tactile—the play between shiny, smooth skin and slightly raised scars. This contrast is not an end in itself; it expresses an important ideal of balance. For the Yoruba, the body and mind are one. The outer reveals the inner, and vice versa. In this way, scarification expresses something genuine about a person's character. It allows one's inner beauty to come through to the surface.

Contemporary Yoruba scarification incorporates current designs, such as airplanes, wristwatches, and scissors. Some even have the name of a loved one scarred upon them, which is comparable to the Western practice of having the names of loved ones tattooed upon one's body.

Multivalence

For most cultures that practice scarification, including those described above, the marks and practice are multivalent. The Tabwa of the present-day Democratic Republic of Congo (formerly Zaire) cover themselves from head to toe with scarification with a variety of types of scars. Scarification is related to the process of creating a social order in a natural world considered chaotic. As such, Tabwa scarification communicates a worldview. Rather than being reduced to merely a matter of beauty, ritual, or symbol, Tabwa scars are considered to be expressions of a lived religion wherein marks are not always dogmatically consistent in meaning or easily reduced to one concept or another, but are often tied to emotion and experience and vary in meaning.

Other cultures not covered in detail here that practice scarification are the Hemba (northwest neighbors of the Tabwa); both sexes of the Bangwa of Cameroon, for purposes of rites and medicinal therapy; both sexes of the Baule of Ivory Coast, for medicinal and aesthetic purposes, as well as to identify themselves as civilized; the Bateke of central Africa, whose facial scarification is said to rival the facial tattoos of the Maori in terms of intricacy, for aesthetic reasons; the Dinka of Sudan, also known for their facial scarification; the Kabre of Togo, as a rite of passage for boys; the Bini of Benin, for purposes of healing and purifying the body and as a rite; the Yombe of the Democratic Republic of Congo, to indicate status and culture; the Ibo of Nigeria, for purposes of tribal identity; and the Pitjanjara of Australia, during dreamtime ceremonies—performances in song and mime of tribal history. Some aboriginal Australians use branding to scar the skin for aesthetic purposes using a hot, glowing stick on the chest, abdomen, back, and arms.

Contemporary Scarification

Reasons for Decline in Traditional Societies

Three major factors influenced the decline of scarification in traditional societies throughout the twentieth century. The first is government intervention (both colonial and nationalist). For example, in the 1970s, the government of Ivory Coast considered "tribal marks" anti-patriotic because they emphasized tribal rather than national affiliation; in addition, these marks were thought to be too primitive. The scarification practices of the Ga'anda were made illegal by the local government in an attempt to promote national unity.

The decline of Tabwa practices was due in part to Catholic missionaries, in particular the society known as the White Fathers, who settled among the Tabwa in the late nineteenth century intending to create a Christian kingdom in the heart of Africa. The White Fathers began the process of suppressing a whole system for conveying meaning and creating order in a chaotic world. Some Christian missionaries thought the practice of scarification was cruel and abusive and viewed it in much the way that contemporary women's rights groups view female circumcision.

Generally, by the late twentieth century, youth from traditional societies wanted to assimilate with urban culture and did not want to be visibly marked with tribal scars. Such scars are considered backward and could lessen their opportunity in the cities. The adoption of Western-style clothing has also hampered the practice. In addition, some youth simply perceive the practice as too painful.

Scars in the West

Scars in Western society have often had a masculinizing effect on appearance. In the late nineteenth century and early twentieth century, this was seen among German university students. Students would get their faces slashed during dueling matches; the resulting scar was a badge of masculinity. Sometimes, students would even rub wine in these wounds to make them more pronounced. This dynamic can also be seen in gang or prison culture, where a scar is proof of strength.

Scars became symbols of rebellion in early punk culture; they also grew in popularity in the San Francisco S/M culture of the 1980s. The modern primitives of the late 1980s and 1990s helped to popularize scarification in Western youth cultures.

Modern Primitives and the Body-Modification Subculture

As scarification is dying out in non-Western traditional societies, it is becoming more common in contemporary modern primitive and body-modification subcultures, which have popularized the appropriation of indigenous body practices, which they see as spiritual. Ironically, as youth from traditional societies leave behind ritual, modern primitives and members of body-modification subcultures incorporate scarification practices in their personal rituals.

Fakir Musafar, often called the father of modern primitivism, is responsible for much of the popularization of scarification as an accepted form of body art. The extensive interviews with him and body piercer/scarification artist Raelyn Gallina in RE/Search's 1989 publication *Modern Primitives* helped to give a meaningful narrative to several types of subcultural body-modification practices.

Concepts that accompany the practice of scarification in traditional societies continue in Westernized renditions of scarification. These include tests of strength, expressing one's inner life or personality, beauty, and even healing. Those who seek out ritual and the therapeutic are sometimes more focused on the possibilities of their experience than on the aesthetic result of the scarification. Those who are interested in the design itself may be more interested in what the scar represents, such as a memorial for a lost loved one. A scar design may also symbolize other deeply held personal beliefs or a belief system.

Methods of Contemporary Scarification

In modern scarring practices, cutting the skin remains a common method. It is also common to go to a professional body-modification artist. The most popular contemporary cutting tool is a scalpel. Various substances such as ash or ink can be rubbed into the cutting in order to achieve a more pronounced scar. Another method used to achieve a more pronounced scar is skin removal. This uses a scalpel to cut the skin and small clamps to help remove the skin. Toro, a Spanish body-modification artist, is thought to be the originator of this method, while artists such as Brian Decker and Ryan Oullette are considered innovators and developers of this technique.

Branding is a common method of scarification. From ancient to modern times, slaves and criminals have been branded for purposes of ownership and stigmatization. Present-day branding takes a different form and is chosen rather than forced. It is popular on college campuses among members of fraternities, especially black fraternity brothers. The professional athlete Michael Jordan has a fraternity brand on his chest.

Branding is also a form of scarification used by modern primitives. Modern primitives often go to body-modification artists with professional methods for branding. One method open for use by scarification artists is the strike method of branding. The artist uses a small bar of some type of metal that is heated until glowing red before being briefly touched to the skin. Other devices are used for branding when greater detail is necessary. Three common devices are cautery pens, electrocautery devices, and electrosurgery devices. All of these devices are able to create more intricate designs than the strike method; however, they vary in their degree of reliability, visible scarring, and intricacy. The electrosurgery device sends an electrical current directly into the skin and allows for the most intricate branding designs. Other methods of scarring include fire direct (adhering a small piece of incense to the skin); tattoo gun scarification (used without ink); Dremel-tool scarification (using a rotary tool with an abrasive attachment); and the least common methods, chemical and injection scarification.

Further Reading: Body Modification E-zine Encyclopedia, http://wiki.bmezine.com/index.php (accessed May 2007); *Body Modification E-zine* Web Site, http://www.bmezine.com; Brain, Robert. *Decorated Body.* New York: Harper and Row, 1979; Faris, James C. *Nuba Personal Art.* London: Duckworth, 1972; Favazza, Armando R. *Bodies under Siege: Self-mutilation and Body Modification in Culture and Psychiatry.* Baltimore, MD: The Johns Hopkins University Press, 1996; Lutkehaus, Nancy C., and Paul B. Roscoe, eds. *Gender Rituals: Female Initiation in Melanesia.* New York: Routledge, 1995; Peregrine, Peter N., and Melvin Ember, eds. *Encyclopedia of Prehistory, Vols. 3 and 7.* New York: Plenum, 2001, 2002; Polhemus, Ted. *Hot Bodies, Cool Styles.* London: Thames & Hudson, 2004; Rubin, Arnold, ed. *Marks of Civilization.* Los Angeles: Museum of Cultural History, UCLA, 1988; Spindler, Konrad. *Man in the Ice.* New York: Harmony Books, 1994; Vale, V., and Andrea Juno, eds. *Modern Primitives.* San Francisco: Re/Search, 1989.

Jaime Wright

Skin Lightening

Skin lightening (also known as skin whitening, skin bleaching, or skin depigmentation) is the process by which the skin is chemically made to appear lighter in color, either by breaking down existing melanin stores or by blocking further production of melanin. Skin lightening first appeared as a means to manage a variety of skin disorders that affect pigmentation. However, given the near-universal prevalence of skin color-based racism and its attendant privileging of lighter skin, skin-lightening methods were quickly adopted by lay populations for their own use. While many legal commercial skin lighteners are currently available in the United States and around the world, many people resort to harsh chemical mixtures or compounds of unknown composition to bleach their skin. While these products do produce a lightening effect, various—often severe—skin and other health-related problems may result from their use.

Dermatological Origins of Skin Lightening

Skin lightening or bleaching was a technique first widely used by dermatologists to assist in the management of chloasmas, melasmas, and solar lentigines—all conditions that can lead to hyperpigmentation, or a marked, often irregular darkening of facial or bodily skin. While hyperpigmentation can be caused by disorders or overexposure to the sun, skin darkening can occur in response to rashes, acne, or injuries—especially among darker-skinned persons. Pharmaceutical skin-lightening techniques are also used in the management of vitiligo (the disease from which pop-music icon Michael Jackson purportedly suffers), a disease that causes marked, irregular depigmentation in patches of skin throughout the body. When used in the treatment of vitiligo, skin lighteners are applied to the areas of skin that have not been depigmented by the disease (that is, the normally colored skin), helping the skin to achieve a more even-toned appearance.

The Cultural Privileging of Lighter Skin

Since at least the 1960s, widespread use of skin-bleaching products has been documented among lay populations in varying parts of the world. This is particularly true in areas with high concentrations of people of color who have been affected by European colonization or chattel slavery. While many over-the-counter and prescription skin-bleaching products are available in the United States, bleaching seems to be the most popular in countries such as Jamaica, Senegal, Mali, Togo, Zimbabwe, Ghana, Thailand, the Philippines, Japan, and the Indian subcontinent. In notable cases, such as Japan, some countries with predominantly people-of-color populations have a centuries-long history of privileging very pale skin (as is seen with geishas). However, most social scientists attribute the popularity of skin lightening in these areas to the lingering effects of the disparagement of indigenous features in favor of European ones during and directly following periods of slavery or colonization.

The outright denigration of native features was reinforced by the systematically better treatment that lighter-skinned individuals received during slavery and colonization. As a matter of course, Europeans assumed that lighter-skinned people of color had some portion of European ancestry (or were in some way more similar to Europeans), and thus were more "civilized" than their darker-skinned counterparts. Lighter-skinned individuals were more likely to be sent to Europe for formal schooling or selected for higher paying jobs that involved working closely with Europeans. During chattel slavery, the lighter-skinned slaves were often the master's or overseer's children. As such, these children received better treatment and were afforded easier tasks—often in the master's house. The more tedious, backbreaking outdoor labor was left to the darker-skinned slaves (giving rise to what some have called the "house slave/field slave" dichotomy). Through time, lighter skin became equated with better treatment and a better life, both for whites and people of color alike.

Although chattel slavery and the era of European colonization are formally over, lighter-skinned people of color continue to fare better than their darker-skinned counterparts. Multiple studies have shown that fairer-skinned people of color (especially women) tend to score higher on multiple measures of well being—including income, employment status, and some physical/mental health outcomes—than dark-skinned individuals. It has also been demonstrated that popular media tend to systematically privilege lighter skin. It would come as no surprise, then, that numerous studies show that at-home skin bleachers tend to cite living up to beauty ideals, access to better jobs, and higher pay among their chief reasons for continuing to bleach their skin.

Skin-Lightening Methods

Although several chemical compounds have been shown to have some skin-lightening properties, the majority of commercially produced skin-lightening products contain one of three major components (or some combination thereof). The most widely used skin-lightening compound is hydroquinone, which inhibits the production of melanin, the pigment that gives skin its color. It has been shown to be a strong irritant when used at higher concentrations, especially when combined with other bleaching agents. Corticosteroids are another popular bleaching agent. Corticosteroids are a class of anti-inflammatory drugs that, in sufficient concentration, have been shown to lighten skin. However, with prolonged use, corticosteroids can thin the skin and affect its ability to repair itself. Mercury is another element commonly found in skin-bleaching products. Mercury is a heavy metal and is highly toxic. Its use in body products is illegal in most countries, but it frequently appears as an ingredient in skin-lightening products sold on the black market.

Because the action of creams and topical solutions is not permanent, most bleachers must continually apply bleaching agents to their bodies over a period of years to achieve some sustained effect. Skin-lightening techniques using lasers afford more long-term solutions. Laser treatment, however, is best suited for smaller areas of skin and is not practical for at-home use.

In addition to the many commercially available prescription and over-the-counter skin-lightening products, homemade and black market skin-lightening products and methods abound. These can range from the use of household or industrial bleaches, detergents, and a variety of other chemicals, including vaginal contraceptive tablets in some cases. Many products available on the black market are assumed to be quite dangerous, as their composition is generally unknown.

Health Risks

Many complications can arise from the prolonged use of skin-bleaching agents, ranging from moderate to severe. According to a variety of clinical data, acne, eczema, and contact dermatitis are among some of the more frequently occurring negative side effects. Some of the more severe reported side effects include dermatophyte infections, acute scabies infestations, and, with certain chemicals, even various forms of cancer. Due to the widespread use of skin-bleaching agents and the high rates of resulting health problems, many countries have declared skin bleaching a major public health problem and have enacted campaigns to end it. *See also* Skin: Cultural History of Skin.

Further Reading: Hunter, Margaret. "'If You're Light You're Alright': Light Skin Color as Social Captial for Women of Color." *Gender and Society*, vol. 16, no. 2 (2002): 175–193. Johnson, Sonali. "The Pot Calling the Kettle Black? Gender-Specific Health Dimensions of

Color Prejudice in India." *Journal of Health Management*, vol. 4, no. 2 (2002): 215–227. Mahe, A., F. Ly, G. Aymard, and J. M. Dangou. "Skin Diseases Associated with the Cosmetic Use of Bleaching Products in Women from Dakar, Senegal." *British Journal of Dermatology*, vol. 148 (2003): 493–500; Persaud, Walter. "Gender, Race and Global Modernity: A Perspective for Thailand." *Globalizations*, vol. 24, no 2 (September 2005): 210–227; Steiner, Markus, John Attafua, John Stancack, and Tara Nutley. "Where Have All the Foaming Vaginal Tablets Gone?: Program Statistics and User Dynamics in Ghana." *International Family Planning Perspectives*, vol. 24, no. 2 (1998): pp 91–92.

Alena J. Singleton

Stretch Marks

"Stretch mark" is a common term used to refer to a benign skin condition in which discoloration occurs on the skin surface due to dramatic changes in body size and shape over a relatively short period of time. Known medically as *striae distensae*, stretch marks usually appear as purple, pink, reddish, or white lines found anywhere on body surfaces, particularly the abdomen, thighs, breasts, and buttocks. Stretch marks are experienced by a majority of people, 50 to 90 percent of the population by some estimates, though women are more often affected. Stretch marks are the physical side effects of rapid changes in body size and shape, including pregnancy, weight gain or weight loss, growth spurts during adolescence, body building, and also conditions that involve hormonal changes. However, these circumstances are not predictive of the development of stretch marks, since genetics, family predisposition, race/ethnicity, and age also play a role in determining their prevalence.

The physical appearance of stretch marks occurs as the collagen in the skin is separated with the expansion (or decrease) of the flesh underneath the skin and a type of scarring occurs as a result of the healing process when the collagen is replaced over a period of time. At first, stretch marks can begin as pinkish-purple tear and gradually lose their color. Eventually, the stretch marks will appear to be smooth, flat white, or silver lines in the skin. Once they do occur, stretch marks are considered to be permanent, though their appearance does fade over time. There is no proven preventative treatment.

The development of stretch marks is particularly associated with pregnancy, which is when many women first experience them. Though there are many changes to the skin during pregnancy, stretch marks are often referred to as a lasting side effect that remains on a woman's body after she has given birth. Historically, it is not clear when stretch marks came to be regarded as a negative side effect of pregnancy, though its "problem" status may be associated with the perception that the skin has been permanently altered or scarred after pregnancy, and that the body then exhibits these visible reminders that a woman has had a baby long after she has given birth. These marks of pregnancy, particularly on the breasts and abdomen, can also be considered a significant problem in a society that views the pregnant body as a temporary phase and puts pressure on women to return to their presumably thinner, smoother, pre-pregnancy forms. The fact that images of models and celebrities (particularly those who have been pregnant) do not display any stretch marks as a result of photo retouching can also contribute to the perception that they should not exist on a desirable woman's body. The concern and self-consciousness about pregnancy-related stretch marks can also be seen as a result of a culture that increasingly puts responsibility on women to prevent or hold off any natural changes or alterations to the skin, from wrinkles to sun damage. Indeed, even though stretch marks are a common result of pregnancy for most women and are not painful, many women will make efforts to prevent them. Unfortunately, there is no established treatment for their prevention, and the severity of their appearance mostly depends on individual predisposition, previous pregnancies, and weight gain during pregnancy.

Practices Used to Prevent and Treat Stretch Marks

Stretch-mark prevention is a common topic that is often advised in books, magazine articles, and Web sites about pregnancy. Advice on how to help prevent stretch marks is often to keep the skin supple, soft, and smooth in order to ease the skin's "stretching" process. This is said to be accomplished with the aid of creams and lotions that are specialized for stretch marks. Common ingredients are cocoa butter, vitamin E, lanolin, and collagen. These creams are to be applied on a woman's abdomen and breasts throughout pregnancy. Castor oil, seaweed wraps, and homeopathic creams and oils have also been suggested as topical treatments to prevent stretch marks from developing during pregnancy. Though there are many creams marketed to pregnant women to prevent stretch marks, medical literature shows that no successful preventative treatments have been found.

Since there are no known ways to completely prevent stretch marks from appearing during and after pregnancy, women who are concerned about their appearance have a variety of options to try to remove or at least minimize them. But stretch marks are often resistant to treatment, and the procedures themselves can be expensive, painful, or both. Contemporary methods to treat stretch marks include different topical regimens with creams that contain acids like glycolic acid or L-ascorbic acid, which break down the outer layers of the skin to improve its appearance. Cosmetic laser treatments are another contemporary way that stretch marks are treated. Administered by specially trained technicians on an outpatient basis, a laser beam is concentrated on the area with the stretch marks; after several treatments, their appearance can be dramatically lightened. However, cosmetic laser treatments are expensive and focus only on small areas of skin. An easier, though less-permanent approach is to cover stretch marks with specially formulated makeup that blends in with the skin. Though stretch marks are a natural result of the many changes that the body goes through during pregnancy, culturally they remain an unwanted, permanent reminder of pregnancy on a woman's body.

Martine Hackett

Subdermal Implants

Subdermal implants are objects that are inserted under the skin in order to change the visual appearance of the body's surface. They vary in intricacy and size, ranging from simple rods or beads to more elaborate carved shapes, and are commonly implanted in various parts of the body, including the head, arms, and chest. Those acquiring implants are a small subset of the wider body-modification community, namely people generally already involved in more prosaic body-modification practices, such as piercing or tattooing, and are looking to further explore the possibilities of their bodies. In this context, subdermal implants are specifically designed to alter the shape of the body beyond normative ideals; while the technologies and techniques for subdermal implantation used in a subcultural framework appropriate much from the processes and procedures of mainstream cosmetic and reconstructive surgery, they share little in terms of history or intention with surgeries such as breast or chin implants, which are designed to render the client more normatively attractive.

The implantation of objects under the skin has historical precedence. Most prominently, members of the Japanese Yakuza crime gangs have been documented to have placed beads or pearls under the skin of their penises, ostensibly with one bead implanted to mark each successive year spent in penal custody but also as a perceived method of increasing sexual prowess. Using a sharpened chopstick or other suitable implement, they pierce a sizable opening into the skin of the shaft of the penis and insert a spherical bead into the resultant wound, which is then bandaged and allowed to heal. The procedure encapsulates the implant under the skin and creates a bumped, lumpy appearance.

The penile beads of the Yakuza were documented in V. Vale and Andrea Juno's *Modern Primitives* in 1989, and this led to experimentation with similar procedures among the nascent body-modification subculture of the early 1990s. At the same time, French artist Orlan (1947–) began to perform her series of theatrically staged cosmetic surgery operations, entitled *The Resurrection of Saint Orlan*, in which the geometry of her facial features was radically reconfigured by the use of subdermal implants under her temples and brow. While Orlan has been clear in distancing herself from the body-modification subculture of the time, her operations being carried out in a managed artistic setting by qualified and registered medical professionals, the radical aesthetic she presented did pique the interest of those already looking beyond piercing and tattooing for ways of exploring the possibilities of corporeal transformation.

Initial experiments took the methodology of Yakuza beading and deployed them to implant similar small, spherical, or rod-shaped plastic or steel objects into parts of the body other than the genitals. By 1996, Phoenix-based body-modification pioneer Steve Haworth had implanted a large, specially adapted piece of body jewelry under the skin of his right wrist, establishing a whole new genre of modification: three-dimensional art implants. The procedure Haworth developed essentially medicalized and formalized the Yakuza prison technique: the skin is cut with a scalpel, the layers separated with a blunt tool known as a dermal elevator to create a pocket, the implant inserted, and the wound sealed by suturing it closed. He realized that there was no need to restrict either the size or shape of the implant, and through word-of-mouth and dissemination of photographs in magazines, Internet newsgroups, and Web sites such as www.bmezine.com, clients from around the world began traveling to Haworth to have implants performed.

From this initial procedure, Haworth and those inspired by him developed a number of different materials such as medical implant-grade steels, silicone, and Teflon that could be carved or molded into any desired form to be put under clients' skin. Nevertheless, simple shapes are the most effective visually, as the healing pocket will generally not create enough definition around the implant itself if the design is too ornate or intricate. Because of this, the most popular forms seem to be hearts, stars, discs, and domes, the latter being particularly popular for forehead implantation as it produces the look of horns. The size of implants is also somewhat limited by the tensile strength and elasticity of the skin under which it is to be implanted, although much in the same way as piercings, implants can be gradually stretched over time if the first implant is removed and replaced by one slightly larger in size.

Implants are generally most successful if placed on top of a fairly firm surface such as the head, outer arm, sternum, or collar. If placed under skin tissue on top of muscle, the implant can sink into the soft tissue underneath and cause a variety of complications, necessitating removal far more difficult than the original insertion. Due to the pseudosurgical nature of these procedures, it has also proven difficult for subcultural practitioners to obtain the highest grades of implantable materials used by professional surgeons, a matter which has embroiled the practice in some controversy both within the subculture and beyond it; some vocal critics even accuse those who perform implants outside of a professional surgical setting of endangering the health and safety of their clients. While implants have not been particularly troublesome in general since their emergence as a popular form of body modification, some particularly experimental uses of implant technologies, including the encapsulation of magnets in silicone sheaths to be implanted under the fingertips, have been problematic and arguably even dangerous.

Further Reading: Ince, Kate. *Orlan: Millennial Female.* Oxford: Berg, 2000; Larrat, Shannon. *Body Modification E-Zine,* http://www.bmezine.com; Larrat, Shannon. *ModCon: The Secret World of Extreme Body Modification.* Toronto: BMEBooks, 2002; Tsunenari, S., T. Idaka, M. Kanda, and Y. Koga. "Self-mutilation. Plastic Spherules in Penile Skin in Yakuza, Japan's Racketeers." *American Journal of Forensic Medical Pathology* 3/2 (September 1981): 203–207; Vale, V., and Juno, Andrea, eds. *Modern Primitives.* San Francisco, CA: Re/Search Publications, 1989.

Matthew Lodder

Tattoos

"Tattoo" refers to a permanent mark, usually of intricate design, created on the body when colored pigment is inserted under the surface of the

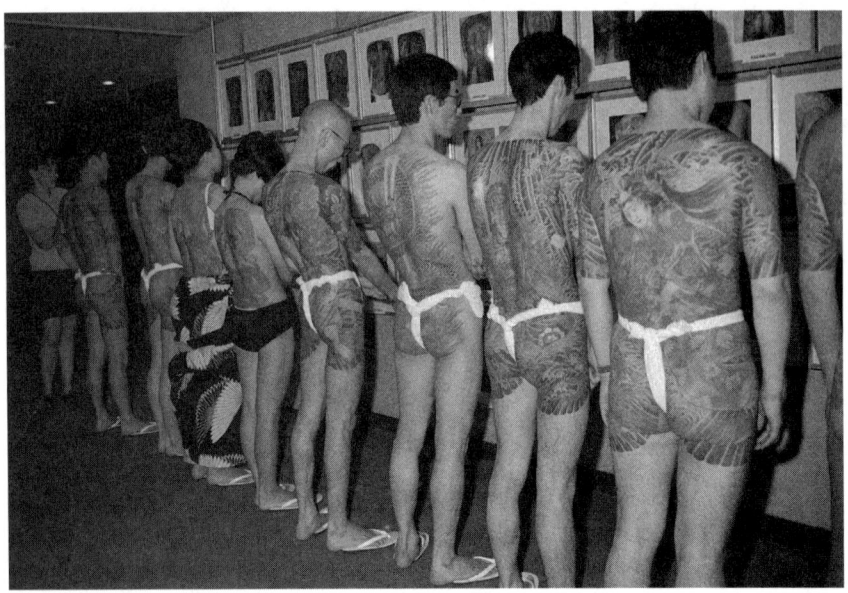

A group lines up to display the work of tattoo artist Tamotsu Horiyoshi during a demonstration of Horimono tattooing at the Foreign Correspondents Club of Japan in Tokyo, 1970. Horimono is a unique Japanese form of tattoo covering most of the body. (AP Photo).

flesh. The practice has an extensive history throughout human cultures. A tattoo is made—usually by a specialist—using sharp instruments which pierce the surface of the skin and impregnate the dye deep enough to avoid its erasure during the normal process of dermatological shedding and renewal. There are a variety of implements used to inscribe tattoos, the most prevalent of which involves the "pricking and inking" of the skin with fine needles, and is the favored method in both contemporary Western practices and traditional cultures in the Pacific region. However, other methods include the use of implements which resemble chisels (Maori) or gouges (Japan); in Inuit culture, tattoos are made by a process of threading ink though the skin, much like sewing.

Tattoos vary in color, size, and design, as well as location on the body. The length of time a tattoo takes to inscribe will vary depending on these features—it may be as quick as a few minutes, or take hours, even days to complete. However, because of the invasive nature of the procedure, a tattoo always involves a degree of pain and blood shed, and a subsequent period of healing to enable the design to seal into the skin.

Because of the painful and bloody nature of tattooing living flesh, as well as the permanent and highly visible alterations to the body it creates, tattooing has, throughout history and across the globe, been afforded a special status in human culture. Indeed, some anthropologists have suggested that it is precisely this kind of practice, which deliberately alters the body in accordance with social values and/or belief systems, that is a primary, defining feature of human culture, and which distinguishes human societies as unique among species. However, the meanings and values embodied by tattoos and their inscription vary hugely: in some cultures, tattooing is a sacred practice, used to demarcate special individuals within a community, or significant phases of life, while in other societies, tattooing is associated with degradation, and has been used specifically to mark those who are designated as aberrations by virtue of their behavior, beliefs, or their "racial," religious, or social status. Elsewhere, such as in Europe, tattooing has a more discontinuous and ambivalent history, oscillating between periods of popularity, as an art form or marks of high social status, to symbols of denigration, stigma, and pathology. In all of these cases, tattoos also have a significant relationship with the specific cultural norms and roles of gender.

Tattoos in Cultural Contexts: History, Methods, and Meanings

Tattooing is an ancient practice. In 2000 BCE in Egypt, it was particularly associated with noble women and those who followed the sacred deity of the goddess Bes. In other parts of the world, including Europe, there have been remnants of tattooed bodies found dating from around 500 BCE. Prehistoric remains discovered in the Alps also indicate the early use of tattooing, possibly as a healing practice—the tattoos covered body parts that, under X-rays, revealed significant bone degeneration and related pain and immobility.

Historically, tattooing was particularly prevalent in cultures around Polynesia and the Pacific area, and the word (and practice) used in the English language today stems from this area. The word "tatu," meaning "to strike," is derived from the specific method of inscription, in which very fine, sharp, pointed, needle-like instruments were tapped into the skin to impregnate dye under its surface. The tattoos would take many days to inscribe and were done by specialists, in some areas accompanied by ritual songs or drumming in order to both enhance the rhythm of the tattooing process, as well as to help the recipient transcend the pain. Both men and women were tattooed in these cultures, although there were fundamental differences in the designs, locations, and meanings of tattoos for each gender.

Likewise, among the Maori people of New Zealand, tattooing is a key practice with deep social, spiritual, and gender dimensions. Maori tattoos are made up of black pigment and clear, solid, spiraling lines, and are inscribed when a young person is of age and competence to fulfill his or her adult role and responsibilities. Once a young man is fully versed in his ritual warrior and social roles, he will take on tattoos that cover his face—called the moko—and the lower part of his body. The intricate spiraling patterns are unique and carefully follow the contours of the face and body in designs that are both symbolically and aesthetically crafted. Thus, each moko is as individual as a fingerprint or signature and, indeed, recreating the design was used in lieu of such during the land transactions between the Maori and the European (*pakeha*) settlers. The tattoos are created using a chisel-like instrument, which creates both a tactile and visual effect and also heightens the physical and emotional intensity of the experience, as well as the endurance required to take on the marks. For centuries, the moko has been a source of fascination and revulsion to Europeans and, until recently, the tattooed heads of Maori chiefs were preserved as museum artifacts, only being returned for burial this century.

Maori females also take on symbolic tattooing when they reach the age of maturity. For women, however, tattooing is less extensive, and focuses around the lips and chin. The designs, also of intricate spiraling motifs, symbolize not only their role in the culture, but also trace the line of female ancestry, marking out a matrilineal aspect of Maori culture and the role of women within it.

There are many other parts of the world where tattooing denotes a specific female role and practice. For example, in many areas of the Middle East, tattooing is an ancient rite, carried out by specialist female practitioners who inscribe the hands and faces of other women. Indeed, across the globe, tattooing appears to be inescapably gendered, in terms of the design and extent of the markings, the places of the body that are decorated, and the roles and rituals they symbolize and embody. Like other practices which mark the body (such as scarification and piercing), for women, tattoos primarily denote their adult female status in terms of their sexual and reproductive role, and their eligibility as wives and mothers (as well as making them attractive to men); for men, the marks depict their social role and status, and their prowess as warriors, hunters, or leaders. The process of inscription and the pain and blood shed also has specifically gendered interpretations; for females, the marks demonstrate their ability to withstand childbirth; for men, the pain is associated with the physical demands of a warrior or hunter. Where tattoos indicate a special spiritual status or role, including many Native American cultures, the process of the tattooing itself, the pain experienced and the blood shed, is also associated with otherworldliness, divinity, and connection with sacred and spiritual dimensions.

In many cultures, following the inscription of a tattoo, a period of taboo is prescribed, in which the newly adorned initiate is kept apart from the community's day-to-day activities. This not only enables healing but also adds further symbolic distinction to the importance of this ritual marking.

In many places, then, tattooing is an ancient cultural practice that has clear social, symbolic, spiritual, and ritual elements that are consistent throughout time and, indeed, form coherent aspects of that culture and its roles and rituals. However, in other parts of the world, the meaning and location of tattooing is less consistent, and fluctuates over time.

In Japan, tattooing has been practiced for millennia, with some estimates dating it as far back as 10,000 BCE. Like other traditional practices, Japanese tattooing has its own specific design styles and methods of inscription. The word "irezumi," meaning to "insert ink," describes a procedure in which ink is applied and then worked into the skin using a bundle of fine, hand-held, gouge-like instruments. The technique is renowned for its capacity to produce extremely fine work, with intricate shading and detailed design in rich colors. Traditional images include flowers, mythological creatures, and characters, and cover the entire body. Indeed, the movement of the body is often incorporated into the design, so that the image itself appears to come alive as the skin and muscles flex.

Irezumi was initially practiced throughout Japanese social strata, but eventually became associated with specific status markers that impacted the breadth and popularity of its use. For example, when prohibitions prevented anyone other than nobility from wearing ornamentation or elaborate and costly garments, irezumi became a badge of status and wealth among merchant classes. These works of art had such value that they became tradable commodities; an individual could put a down payment on a skin and claim it after death. Particularly fine skins were displayed by collectors in private museums. However, in the nineteenth and twentieth centuries, tattooing became associated with the criminal and lower classes.

For women, the status of the practice equally changed over time, shifting from a revered art form to a mark of denigration. Female tattoos were initially symbolic of ideal feminine sexuality and were associated with geishas, beauty, and femininity, and the tattooed woman was held in high esteem. Delicate designs were used to enhance the erogenous zones, and to cover the entire surface of the body—naked flesh was perceived as neither attractive nor desirable. But, just as male tattooing became associated with undesirable aspects of society, so too did female marks, and eventually they were worn primarily by prostitutes and women of low social status. In recent decades, however, there has been a revival of interest in the traditional practice both within Japan and across Europe and the United States; the status of irezumi as an art of great skill and beauty is once again being recognized and popularized.

Japan is not unique in terms of the shifting status and fluctuations in popularity and practice of tattooing. Across Europe, fragments of evidence indicate that tattooing was practiced among certain ancient cultures and/or tribes. Indeed, the word that was commonly used to denote the people of ancient Britain derives from an expression that describes them as "painted in various colors," though blue pigment was most commonly used. There is also some evidence that the early Celts in both Ireland and Scotland may have also used tattooing, and there are a number of words in the Irish language that have connections with practices of tattooing.

However, first the Roman, and then the Christian era brought massive changes to the social organization and values of the disparate tribes of Northern Europe and, with them, a very specific and hostile response to tattooing. The Romans, like their Greek predecessors, had come into contact with cultures where tattooing was widely practiced. But for them, rather like the European colonizers who followed centuries later, tattoos and other indelible marks on the flesh were signs of barbarism. The Greek word for stigma is closely linked to the word tattoo, and both Greek and Roman cultures used tattooing as marks of denigration which were inflicted upon criminals, slaves, and those about to be executed. Christians were often tattooed in this way by their Roman captors, inscribing a mark of double stigma, in that it both degraded them within the culture of their oppressors, but also contravened biblical direction which forbade Christians to mark their flesh, for in Christian doctrine, the human body is both made by, and in the image of, God, and therefore marking the flesh and altering the body become acts of blasphemy. Nonetheless, and in yet another reflection of a process that was to recur (among other groups) much later in European history, the practice of tattooing among Christian devotees developed a rather more complex and ambiguous status. As the persecution of Christians by the Romans continued, the act of *voluntarily* taking on a mark that depicted their beliefs and commitment to them, in spite of the danger that ensued, became a badge of devotion and faith. To be willing to risk capture, punishment, and death by displaying Christian motifs on the body—often a fish or a cross tattooed on the hand—demonstrated not only an unwavering commitment to the faith itself, and the willingness to die for it, but also a certainty of redemption and salvation in the afterlife. Later still in Christian history, tattoos once again became popular marks of faith, despite their continued prohibition within official doctrines and papal decrees. The early Middle Ages witnessed the rise of pilgrimages as acts of devotion, and these religious travelers often returned from the Holy Land with tattoos to mark both their scared journey and their commitment to their faith.

Nonetheless, tattooing retained a primarily negative status in relation to European and Christian culture and, indeed, during the witch trials, indelible marks on the skin were seen as evidence of devilment and witchcraft. Thus, despite evidence of knowledge of it, as well as of some degree of cultural presence and practice, tattooing was not widely incorporated into European culture and was largely dormant until the eighteenth century.

In recent history (the eighteenth century onward), tattooing is largely reputed to have been reintroduced to European culture as a result of the voyages of the explorer Captain James Cook (1728–1779)—indeed, he is also credited with bringing the word "tattoo" into the English language. Cook and his contemporaries were fascinated by the body practices they witnessed during their travels, and often took on the ritual inscriptions of the cultures they encountered. However, their openness to this "exotic" practice of permanently coloring the skin with intricate and elaborate designs was not widely shared, and a short-lived European fascination with tattooing was quickly replaced with a more hostile response, reflected in the dominant religious and social attitudes of the day. In European and American cultures, the status of tattooing was largely shaped by the values of colonization, where non-white, non-European, non-Christian societies were/are regarded as savage, primitive, and barbaric, and in need of "civilization." Colonizers and missionaries responded to the body practices that they encountered with fear and revulsion. They saw tattooing—and especially the tattooing of women—as barbaric mutilations which went against the laws of god, civilization, and aesthetics. (However, European customs that deformed the female body through practices such as corsetry, or using lead to whiten the skin, were never subject to such scrutiny.) Tattooing was outlawed by many colonial governments, resulting in the loss of traditional practices in many parts of the world, including Polynesia, and explicit prohibitions were frequently reiterated in Christian doctrines. Nonetheless, tattooing has maintained a presence in recent European and American cultural history, albeit in a discontinuous way, and with an often ambivalent or negative status.

Following Captain Cook, the practice of tattooing briefly spread not only among seafarers, but also among other tradesmen and guilds of craftsmen who used tattoos to depict their trade. This had largely practical motivations, especially for journeymen, enabling them to convey their trade where differences in dialect, language, or literacy could inhibit communication. However, tattoos primarily existed only among the lower classes, and these "barbaric" practices from the "new worlds" were generally regarded as marks of degradation. Indeed, in Europe, branding and tattooing were adopted as means of specifically marking criminals and slaves; for example, in parts of Russia, facial tattoos were inscribed upon convicts. However, it was not long before this practice became subverted and, much like the Christians of previous history, convicts began to inscribe themselves with their own designs and marks, reclaiming and reidentifying the marks of stigma. These tattoos soon developed into elaborate codes, which identified membership in particular gangs and status within the criminal community. These badges of honor became of great concern to the authorities, who, excluded from the new subculture, and threatened by the implications of its development, were anxious to understand and decode their meanings. These self-inflicted marks also became the focus of many psychomedical theories that regarded the tendency to mark the body as inherently connected to innate predispositions toward criminality and various forms of social and psychopathology.

The late nineteenth century saw a brief reversal of this trend, and evidence of possibly the first "tattoo renaissance." Technological developments, most notably the rotary mechanism and the electronic sewing machine, enabled new, more precise tattooing equipment—very similar to that which is used today—resulting in tattoos that were much finer in detail and design, and much easier to inscribe. A surge of popularity ensued, particularly among the upper classes of Europe, where tattoos quickly came to symbolize worldliness and "art." Professional tattoo studios opened in both London and New York City, and even Queen Victoria and her husband were reputed to have tattoos.

Tattooed bodies also became popular spectacles of circus sideshows and freak shows. The tattooed individual—initially men, but later women—would display his body, entirely covered with tattoos, and recount a tale of capture and enforced inscription by some "exotic" tribe. However, for women, the display of the body was counter to the sexual mores of the time, and so "the tattooed lady" eventually became associated with a form of sexual permissiveness and voyeurism, which reduced her social status. Touring shows more or less disappeared by the beginning of the twentieth century, and tattooing once again slipped into a cultural decline, its existence fragmented, and its practice reverted to use upon, or by, groups of stigmatized people.

Tattooing briefly reemerged in the first half of the twentieth century amid the horrors of war and genocide where it inscribed the brutal marking out of the socially outcast. During the Holocaust, many victims of the Nazis in concentration camps were tattooed with a serial number on their forearm. This mark was of manifold degradation, not only dehumanizing the wearer by reducing him or her to the number inscribed, and the abuses inflicted upon his or her body, but also because tattooing explicitly contravenes the teachings of the Torah. Many Holocaust survivors who bore the marks retained a complex and ambivalent relationship with them. Some had them removed, while others retained them as evidence of the horrors they had endured, hoping the tattoos would maintain their visibility in the larger cultural memory.

During the early twentieth century, tattooing also maintained a presence, albeit a highly stigmatized one, among certain groups of working-class men. Favored designs included images of birds (usually a swift or swallow), a cross or crucifix, a heart with a scroll containing a name or "mom and dad," or sexualized images of women. Tattooing maintained its popularity in penal institutions, and self-tattooing among convicted prisoners, as well as incarcerated young people, was commonplace, and remains so today. Prison tattoos are often coded. For males, a dot on each knuckle means "all cops must die" or something similar; for females, a dot below the eye, known as "the borstal spot," is popular.

Tattooing in Contemporary Europe and the United States

The stigmatization of the tattoo is also connected to its rise in popularity among subcultures in Europe and the United States from the middle of the twentieth century onward. The late 1950s and early 1960s witnessed the emergence of a number of youth subcultures across Europe, North America, and the English-speaking world. Changing social mores and economic booms facilitated the growth of these subcultures. Adherents expressed alternative values and interests through music, clothes, and, very soon afterward, the reemergence of tattoos. One of the first subcultures to embrace tattooing was bikers and, most notoriously among them, the Hells Angels. Bikers and Hells Angels identified predominantly as outsiders or outlaws to mainstream society and developed extensive countercultural systems based on a kind of hypermasculinity. Different tattoos would not only denote membership in a particular gang, but also the stages of initiation the person had completed.

Other subcultures that emerged during this period had rather different values but, nonetheless, body politics played an important role. The hippies, who were equally opposed to mainstream values, but who embraced mottos of love and peace, began to challenge some of the gender and sexual norms of society: men grew their hair long and adorned themselves with beads and clothes made of light, floral, and "feminine" fabrics; women discarded makeup, high heels, and bras, and even went topless at hippie gatherings. Body painting in floral, psychedelic, and colorful designs was popular, and for some this extended to tattooing. However, it was among a subculture that emerged shortly after the hippies, and that maintained a contentious relationship with them, that tattooing became one of the primary and most public badges of notoriety and youth rebellion.

The punk movement embraced some of the most visual and total rejections of the norms of mainstream culture. Punk men and women aimed to shock, to publicly and visibly transgress the norms of "civilized society," and to reject everything mainstream. Punks wore ripped, often homemade ensembles fashioned from various clothing and other items, dyed their hair, spiked and shaved it into elaborate shapes, wore studded belts and collars, and placed safety pins through piercings in their noses, ears, and cheeks. Female punks made deliberately ironic combinations of highly sexualized items of feminine apparel; fishnet stockings would be ripped and worn with tiny miniskirts, jackboots, and black lipstick. They, like their male counterparts, also spat, swore, and vomited in public and adopted aggressive and unfeminine body language and behavior. Tattooing quickly became an integral part of punk culture, and both men and women had the names and icons of their favorite punk bands or other countercultural mottos and images inked onto their flesh. Homemade tattoos were also fairly popular; because of their often-misshapen and blotchy appearance, their literal defacing of the skin, they provoked even more hostility than professionally inscribed versions. Indeed, reconnections between what were seen as forms of self-mutilation and tendencies to social and psychological pathology, deviance, and criminality, resurfaced in both medical and public discourses around the punk movement. However, the form of tattooing which caused most offense was the use of Nazi symbols, and in particular the swastika. For some punks, these highly emotive and politically problematic emblems were simply an ironic or shocking gesture, but the tragedy of recent history made them utterly intolerable for many in mainstream society. Further, the burgeoning movement of neo-Nazis across Europe—the skinheads—who also wore ripped jeans, jackboots, and swastika tattoos, and who carried out organized racist and anti-Semitic attacks on property and violent assaults on individuals, blurred the boundary between ironic rebellion and virulent anti-Semitism.

Other groups also adopted tattooing as mark of their particular status, and tattooing became widely practiced among followers of heavy-metal bands. Indeed, the rock stars of the 1970s began this trend by sporting extensive tattooing themselves. However, with the exception of the punks, these new cultures of tattooing were still highly gendered, and social norms of male and female behavior and appearance made tattooing much more acceptable and widely practiced among males, especially when the tattooing was extensive and highly visible. Further, despite this subcultural resurgence in popularity of tattooing, and the increasing prevalence of tattoo studios, it still maintained a fairly low status in mainstream society, associated with subculture, and in particular with deviant, dangerous, and undesirable forms of masculinity.

However, the late 1980s onward witnessed a broader shift in social attitudes toward tattooing, and a professionalization of both its practice, status, and client group has occurred. This shift in cultural attitudes toward the tattoo, known as the tattoo renaissance, affected many areas of society that had previously remained immune to the appeal of tattoos. Mainstream cultural icons and their followers have become devotees, and models, pop stars, actors, and sports figures have increasingly displayed tattoos on their bodies. Designer tattoos have become a mark of status among materially successful and socially conformist people from middle-class backgrounds. However, these tattoos, in design, location, and extent remain highly gendered, and female tattoos closely follow the norms of femininity. They are generally discrete and primarily located on areas of the body which are both easy to conceal and which are also highly sexualized, including the hips, buttocks, breasts, and lower abdomen. For men, so-called tribal tattoos that mimic traditional styles of ancient cultures and often cover the arms are popular, as are large "back pieces" made up of a single image or design that covers the entire back and takes many sittings to complete.

Homemade tattoos as well as those associated with traditional working-class culture maintain a degree of ambivalence or stigma, as do tattoos on certain parts of the body, including the hands, neck, and face. Likewise, very extensive tattooing, particularly when combined with other forms of body modification (including piercing and branding), is still not widely acceptable in European and U.S. culture and, as such, remains integral to contemporary subcultures.

Two of the most recent countercultural movements that have reinstigated the debates regarding the relationship between forms of body modification, deviance, and "self-mutilation" have emerged in parts of the United States and Europe. These two groups, the modern primitives and the queer movement (which refers to gender and sexual orientation and includes lesbian, gay, bisexual, transgendered, and transsexual [LGBT] individuals, as well as those who practice sadomasochism), have incorporated tattooing and other forms of body modification into their identities.

The modern-primitives movement was founded by Fakir Musafar (1930–), who first used the term in 1978 (it came into more common usage during the 1980s). The ethos of the modern primitives is based upon a rejection of contemporary Western values, especially the reification of the mind over the body but also the individualist and ecologically destructive aspects of modern capitalism. Instead, modern primitives embrace the spirituality, cultural values, and practices of ancient, small-scale societies. These cultures are reified for their spiritual relationship with nature and the environment, their alleged social inclusiveness—and in particular in terms of gender and sexual equality—and the ways in which the body was integral to both social and spiritual practice. The modern primitives view themselves as a tribe apart from mainstream society, and as a revolutionary challenge to the values and practices of dominating, globalizing Western culture. They not only adopt extensive tattooing, in which they recreate both the symbols and the rituals of non-Western societies, but also engage in a plethora of practices from indigenous cultures including branding, earlobe extension, and other rites of passage that alter the body. In doing so, they attempt to challenge contemporary cultural norms regarding the relationship between pain, the body, and wisdom. Modern primitives argue that they experience the ritualized, painful, and bloody practices as processes of deep self-knowledge, spirituality, transformation, and growth—akin to the cultures from which they originated. Modern primitives aim to create an alternative, visibly distinctive, and politically radical counterculture in which body modification is a social, spiritual, and political form of expression.

The other social movement that shares many features with the modern primitives is the queer movement. Its proponents likewise visibly challenge the norms and values of mainstream society and adopt the ritual practices and belief systems of other cultures as a radical expression of identity. Queer subcultures not only reject the mainstream norms of the body, sex, gender, and sexuality, but also embrace a vividly sex-positive attitude, which is often articulated through practices such as sadomasochism and public expressions of sexuality, sexual orientation, and identity. Here too, the norms of mainstream society in relation to pain and pleasure are radically challenged, and ritual cutting, branding, and genital piercings and modifications are seen as integral to both sexual practice and identity. This queer movement fundamentally rejects social pressures toward conformity and assimilation, and articulates a clear minority status. Like many of the other visibly distinctive subcultures, the queer movement has been subject to much hostility and pathologizing scrutiny from both mainstream culture as well as from other parts of the LGBT movement. Their "visibly queer" practices concern many critics, some of whom believe that they suggest psychopathology, while others worry that they reenforce the problematic notions of sexual deviance and perversion that LGBT activists have worked so hard to challenge. Still other critics are concerned that the appropriation of non-Western rituals unintentionally reiterates a form of cultural imperialism. Nonetheless, queer subcultures maintain a visible and provocative challenge to mainstream cultural norms of the body, gender, and sexuality.

Overall, then, and not withstanding its recent renaissance, tattooing in contemporary society continues to retain a highly ambivalent and often clearly gendered/sexual status and, at the same time, an inescapable historical and cross-cultural presence and significance.

Further Reading: Brain, Robert. *The Decorated Body.* London: Hutchinson and Co, 1979; Caplan, Jane, ed. *Written on the Body: The Tattoo in European and American History.* London: Reaktion Books, 2000; Ebin, Victoria. *The Body Decorated.* London: Thames & Hudson, 1979; Groning, Karl. *Decorated Skin: A World Survey.* Munich: Frederling and Thaler, 1997; Hewitt, Kim. *Mutilating the Body: Identity in Blood and Ink.* Bowling Green, OH: Popular Press, 1997; Inckle, Kay. *Writing on the Body? Thinking Through Gendered Embodiment and Marked Flesh.* Newcastle-Upon-Tyne: Cambridge Scholars Publishing, 2007; Pitts, Victoria. *In The Flesh: The Cultural Politics of Body Modification.* New York: Palgrave, 2002; Rubin, Arnold, ed. *Marks of Civilization: Artistic Transformations of the Human Body.* Berkeley: University of California Press, 1988; Sullivan, Nikki. *Tattooed Bodies.* Westport, CT: Praeger, 2001; Vale, V., and Andrea Juno. *Modern Primitives.* San Francisco: V Search Books, 1989.

Kay Inckle

Skull

Trephination

Trephination, the drilling of a hole in the skull, often to cure various ailments, was practiced by a range of indigenous cultures. It was one of the earliest attempts of direct invasive surgery. The modern term "trephination" has etymological roots in the Greek term "trypanon," meaning "drill-borer," which was an instrument used to create a series of small holes. The references to trephination in medical and historical literature as well as illustrations in ancient manuscripts have suggested this procedure was used for a variety of reasons. Some were medical. Meningitis, internal bleeding from trauma to the skull, migraine and other severe headaches, osteomyelitis, epilepsy, syphilitic lesions of the cranium, and fissure fractures have all been treated with trephination. Also, trephination was employed for magical, spiritual, or religious reasons, such as enabling evil spirits or demons to escape the head. Trephination was sometimes believed to be able to prolong one's life. It has been used for easier access to the brain in early neurosurgery.

Archaeological evidence of trephination has been found in numerous locations around the world, including the British Isles, China, the Danube Basin, Denmark, Egypt, France, Germany, Hungary, India, Japan, Jericho, New Ireland, Palestine, Peru, Poland, Portugal, Sakkara, Spain, Sudan, and Sweden. An island north of New Guinea, New Ireland, is known to have practiced trephination to help natives live longer. It has also been suggested by numerous sources that near the Caspian Sea, if a wound was due to the violence of another and trephination was necessary to heal the head, the aggressor paid the surgeon's fee.

Trephination first appeared in the Mesolithic Stone Age, with skulls from this period having depressions. Primitive endeavors at trephination date from between 10,000 and 5,000 BCE. Several kinds of people were responsible for this operation, including medicine men, shamans, early spiritual doctors, and surgeons. In the first century CE, Cornelius Celsus applied the writings of Hippocrates in his techniques for trephination as well as amputation and techniques for setting fractures. Paulus of Aegina, a seventh-century

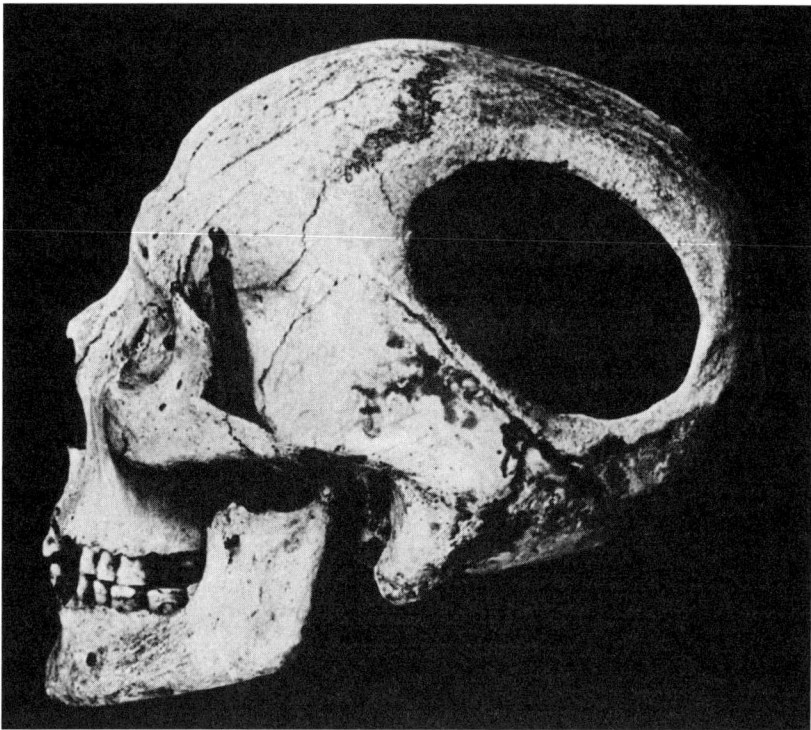
Trepanned Neolithic skull, found at Nogent-Les-Vierges, France. © Bettmann/CORBIS.

Byzantian physician, was one of the first to refer to a unique instrument used in this procedure. The ancient Greek physician Hippocrates was supposed to have used trephination, as did the Indian surgeon Sushruta in the 400s BCE. Doctors in the Renaissance used trephination as a cure for seizures. Modern medical users of trephination include the sixteenth-century French surgeon Ambroise Paré (1510–1590), Sir William Macewen of Glasgow (1848–1924), and Harvey Cushing (1869–1939) of the United States, pioneers of brain surgery.

The survival rate of most patients was very high for this procedure. This is quite astonishing when one considers the methods and instruments used, and the lack of anesthesia in almost every case. Various instruments have been utilized over the centuries. Devices such as pieces of flint, shells, obsidian knives, hammers and chisels, convex scrapers, a surgical modulus, saws, and sharp rocks have all been employed in trephination. Archaeologists have found instruments such as a bronze-crowned trephine, and these artifacts can be used today to learn more about this ancient practice.

Many skulls that have been discovered that show the holes have been made in different ways, with varying sizes and shapes of the area cut into the skull. The holes are most commonly slightly round, but are also triangular and quadrangular. Some skulls have also been trepanned more than once, with two, five, or seven holes. Incan skulls found at Cuzco, for example, have seven boreholes, which were made at different times in the person's life.

South America and Egypt

Egyptian archaeological sites have produced artifacts from the Twelfth Dynasty. In the Anatomy Department of the Cairo School of Medicine is the skull of Princess Horiesnest Meritaten. It has a hole with edges consistent with wounds found in other skulls made with a hammer and chisel. At the Museum of Fine Arts in Boston, computed tomography (CT) scans found a trephine wound in another mummy from the Twelfth Dynasty. Six other skulls from about 1200 to 600 BCE have been discovered in the Sudan and Sakkara, Egypt. The societies may have practiced trephination by scraping with a stone.

Some skulls found and dated to the Mesolithic Stone Age also have circular holes, dating the oldest known efforts of trephination at 10,000 to 5,000 BCE. Archaeologists investigating pre-Incan Peru have found 10,000 well-preserved mummies dating to 2000 BCE, with more than 5 percent showing evidence of trephination. Many of these skulls have multiple skull holes, some as large as seven by eleven centimeters.

It is likely that Andean civilizations such as the Inca used trephination, but it is difficult to sort out the archaeological record of trephination, since skulls were also modified after death as trophies for victorious warriors.

Asia and Eastern Europe

In the second or third century BCE in China, during the Chou Dynasty, surgeon Hua T'o wrote about his medical and surgical practices, including trephination. All of these writings were burned after he died, but legend has made his character a deity of surgery. On one occasion, according to legend, T'o was requested to trephine the skull of Prince Sao, possibly due to headaches. He was ordered executed because the prince thought T'o wanted to murder him.

In Europe, trephination has been recorded in the British Isles and on the continent, including in Eastern Europe. One archaeologist suggested a skull that he found in a prehistoric tomb in central France had been opened to be used as a drinking cup. While this hypothesis was possible, the more likely explanation is that of trephination, as the marks were uniform with those of trepanned skulls using a flint scraper. It is possible that the survivors of these operations may have been thought to have mystical powers, giving their skulls a posthumous value and demand as charms.

Further Reading: Ellis, Harold. *A History of Surgery.* London: Greenwich Medical Media Limited, 2001; Estes, J. Worth. *The Medical Skills of Ancient Egypt.* Canton, MA: Watson Publishing International, 1993; Haeger, Knut. *The Illustrated History of Surgery.* Sweden: Bell Publishing, Suffolk: Harold Starke Publishers, 1989; Jackson, Ralph. *Doctors and Diseases in the Roman Empire.* Norman: University of Oklahoma Press, 1988; Nunn, John F. *Ancient Egyptian Medicine.* London: The British Museum Company Ltd, 1996; Rutkow, Ira M. *Surgery: An Illustrated History.* St. Louis: Mosby-Year Book, 1993; International Trepanation Advocacy Group. http://www.trepan.com.

Margaret Howard

Teeth

Cosmetic Dentistry

For thousands of years worldwide, people have been changing the appearance of their teeth. Teeth have been removed, filed, or replaced. In other cases, they have been dyed, adorned with studs of varying shapes and colors, or covered with different materials. Cosmetic dentistry is the branch of dentistry that improves the appearance rather than the health of the teeth. This definition has important implications as some dentistry can improve appearance at the same time it improves health. In other cases, decisions about whether dentistry is "cosmetic" are determined by the definition of "health" and the cultural notion of "good appearance."

The definition of cosmetic dentistry depends on what "health" is. Traditionally, dental health has been seen as the absence of disease. Increasingly though, broader definitions of health have also considered social and spiritual well being as a resource that allows people to cope within their environment and realize their aspirations. From this perspective, changes to appearance that help one to feel better or get on in life are improving health and are therefore not merely cosmetic. Cosmetic dentistry is also determined by how "good" dental appearance is defined. Most research suggests that the Hollywood smile of straight, white teeth is increasingly pervasive, but there are many historical and contemporary examples in which other presentations of the teeth are highly prized.

An implication of these definitions is that it can be difficult to determine whether an individual case represents cosmetic dentistry or health care. Such debate is not only interesting in relation to anthropological study, but also has practical consequences about which items of dental treatment should be paid for as part of health care. For example, health insurers and national health services such as those in the United Kingdom must decide whether they will fund cosmetic dentistry if it may not improve health.

Dental Transfigurement

Dental transfigurement is the anthropological term usually applied to changes to the appearance of the teeth. It may take the form of replacement of missing teeth, removal of teeth, changes to their shape, or adornment with dyes or artifacts. Contemporary cosmetic dentistry fits this pattern.

One important feature of anthropology is that just because a practice is documented in one setting, does not mean that it was the only setting where it occurred. Likewise, observations from a small number of historical remains from a culture cannot be generalized to include the entire culture of those people. The best-preserved remains are likely to come from the most affluent members of the community (for example, not all ancient Egyptians were mummified in huge tombs). Like many sciences, the study of culture is limited by the evidence available.

There are examples of ancient Egyptian remains with natural teeth held in place by wire. Dental historians disagree about whether these teeth were wired in during the life of the person or whether they were added after death so that the corpse could face the afterlife as whole as possible. More certain examples of teeth being replaced in living people occurred about 500 BCE among the Phoenicians (in present-day Lebanon) and Etruscans (in present-day Italy). In these cases, carved ivory or human teeth were fixed to the remaining natural teeth with wires or bands. It is doubtful whether such teeth could have assisted in the functions of the mouth such as eating or speaking, and so their purpose may have been entirely cosmetic. Pre-Columbian remains have pieces of seashell wedged into the bone to replace missing teeth, and X-ray analyses show that the bone had responded to these false teeth, indicating placement while the person was still alive.

Complete dentures to replace all the teeth became more common in the latter part of the second millennium after the introduction of sugar to European diets. By the eighteenth and nineteenth centuries, dentures were made across Europe and the United States. For example, President George Washington (who had dentures variously made from gold, hippopotamus tusk and elephant ivory, wood, and human teeth) had to have his mouth padded with cotton wool to restore the shape of his face for a portrait. Again, the functional capacity of such treatment appears to have been limited, and if the intention was largely to restore appearance, then it would be deemed cosmetic.

The extraction of one or more teeth has been a custom across many cultures over time. In some cases, the removal of teeth is not for the sake of appearance, but is health-related. For example, the removal of the developing canine tooth ("ebinyo") is still believed by some Ugandan people to prevent or treat childhood fevers. In other cases, two upper incisor teeth were removed among the Dulit Dusun people of Borneo, reportedly so that they could gain a stronger blast with the blowpipes they used for hunting. In other cases, the motive is more cosmetic. The intentional removal of front teeth has been practiced in southern Africa for about 1,500 years. While the "Cape Flats smile" may be declining in popularity, it still appears to be a cosmetic preference among people from some ethnic groups of the Western Cape.

Deliberate modification of the shape of the teeth by chipping and filing is also a common finding in anthropology. For example, the remains of a group of Viking men (ca. 800–1050 BCE) were found to have carefully filed horizontal grooves on their upper incisors. It is not clear whether the grooves indicated a particular occupation or were purely decorative. Filed teeth were common in Borneo until the early twentieth century, and the pattern of filing varied between men and women and between different tribal societies. In some cases, the upper and lower incisor teeth were filed to sharp points so that they interdigitated when brought together. Tooth filing is still a cultural practice in a number of settings, including Ticuana in Brazil, Central Java, and Bali, although the modifications are often limited to notches in along the incisal edge of the upper front teeth.

Many peoples have changed the color of the teeth. Teeth were dyed black among the Mayans (300–900 BCE) of Central America and the Dyaks and Tuaran Dusuns in Borneo as recently as the mid-twentieth century. Although the practice is declining, the Kammu people of Southeast Asia still blacken their teeth with burnt vegetable matter and tar. Interestingly, the pigments used in tooth blackening may protect the teeth from tooth decay, although it is unclear whether this is the primary motive or a side effect.

Some Mayans and people of Borneo also adorned their teeth with pieces of metal or stone. A hole would be made in a tooth with a bow drill and then a carefully shaped stud or inlay would be cemented into the hole. The inlays were of many different colors and shapes, including jadeite, hematite, turquoise, and iron pyrites. In some instances, all the teeth might be covered, as was the case with some Inca, who hammered gold shaped as a veneer over their teeth.

Modern Cosmetic Dentistry

Dental transfiguration in the contemporary developed world is called cosmetic dentistry. Variations of all of the practices documented at other times and places can be found in these settings. In many cases, the intention is to create a real or hyperreal appearance, but there are also examples where the goal is ornamentation with something obviously artificial.

Modern dentistry replaces missing teeth with false ones made of ceramics. The false teeth may be permanently attached to adjacent natural teeth (a bridge), may be clipped into place during the day (a partial denture), or may be permanently attached to titanium implants that are integrated into the jawbone. Such treatments are common in developed countries, and technological developments allow them to look practically indistinguishable from natural teeth.

Where no natural teeth are present, a full denture is held in place by suction and muscle control. In cases where many or all of the teeth are being replaced, there is the scope to create a dentition with a new arrangement and color. Some older people who wear full dentures prefer to have their teeth to be smaller, straighter, and whiter than might be expected in somebody their age, leading to an unnatural appearance, which they obviously prefer for cosmetic reasons.

Orthodontic treatment to move teeth within the mouth (using braces) is usually cosmetic as there are no health consequences of crooked teeth in most cases. The entire removal of teeth for aesthetic purposes occurs but is uncommon. Unsightly natural teeth may occasionally be removed and replaced with a denture, especially for people who prefer not to have long courses of dental treatment. Some models and actresses may also have had their back teeth removed in the belief that this would reduce support for their cheeks, so making their cheekbones more prominent.

Many dental texts describe how teeth that are malformed or misaligned can be reshaped to make them look more typical. When done by dentists, such treatment might involve removing parts of the teeth with a drill or abrasive, possibly with the addition of tooth-colored adhesive materials. Occasionally, teeth are filed to make them look unusual in some way, and since there is no change in function, the effect must be deemed to be cosmetic.

Changing the color of the teeth is now common. Teeth may be bleached whiter using home kits or professionally applied chemicals. For many years, ceramic or metal crowns have been used to restore teeth but are now also used merely to change their appearance if the teeth are misshaped or malpositioned. In most cases, the goal will be a more typical appearance, but even healthy teeth are occasionally modified to look unusual. Gold teeth are found attractive by some people, particularly as this metal is linked with wealth. Gold crowns that fit over the teeth can have a hole cut into them so that the natural tooth shows through, often in an ornate shape such as a heart. Other ostentatious displays of wealth include rock stars who have had diamonds set into their teeth. These practices have been parodied by the Welsh musician Goldie, who has gold teeth inset with diamonds that spell out his name.

Relevance in Contemporary Society

Dental disease was common in most developed countries during the twentieth century, meaning that many teeth needed to be restored or replaced for functional reasons. As dental disease declined in the latter part of the century, the technology that allowed restoration of function is now available for purely cosmetic purposes. Some dentists have seen cosmetic work as an exciting opportunity for an expanded role for dentistry. All these factors make it difficult to decide whether specific procedures are to improve health or are purely cosmetic.

The relevance of cosmetic dentistry in developed countries may go far beyond the condition of the teeth. Notions of attractive teeth seem to be relatively consistent across these settings, with white teeth, with more teeth displayed, and a symmetrical smile regarded as most attractive. Nice teeth are thought to be important for people in prestigious or visible jobs, and those with attractive teeth are judged to be more intelligent, psychologically adjusted, and socially competent. Furthermore, people with damaged and missing teeth show a range of psychological and social impacts from their condition.

However, there is little good quality evidence that dentistry to improve appearance can bring about psychological or social benefit. That is, dentistry may improve appearance per se, but may not enhance health. The teeth comprise only a small part of the face, and the contribution of dental appearance to overall facial appearance may be correspondingly small. In addition, the underlying psychological factors may be stable traits that are not susceptible to changes in physical appearance. People who are self-conscious may never be satisfied with their appearance.

There are broader concerns about cosmetic dentistry that see it as a form of medicalization of beauty and of iatrogenesis (disease caused by doctors). The appearance of minor dental anomalies may have no adverse impact on some people's lives. They may not even be aware of them. Dentists may draw patients' attention to these minor dental defects and suggest they have them corrected. This "supplier-induced demand" has prompted ethical questions. In cultural iatrogenesis, medicine makes disease unnecessary and therefore unbearable. Attractiveness is perhaps more important now than ever before, and society is said to be suffering from a normative discontent with personal appearance. Dentistry may contribute to these problems if the availability of treatment to enhance dental appearance makes deviations from the norm less acceptable. This message is amplified in print and broadcast media that expose people to images that may be enhanced both surgically and electronically. A study of UK teenage magazines found that more than three-quarters of the images had teeth whiter than any restorative materials, and that exposure to these images increases dissatisfaction with appearance. From this perspective, cosmetic dentistry might have adverse effects both on the people who receive it and on wider society; consideration may be required about whether its growth should continue unchecked.

There has always been dentistry to improve the appearance of the teeth. While notions of good dental appearance have changed, the techniques used have remained broadly similar, albeit with enhanced technology. Dentistry can improve dental appearance, but there is little evidence it has broader psychosocial benefits. Dental treatment may affect not only the people who receive it but also those who do not. Consequently there are clinical, ethical, and philosophical limits to the scope of dental treatment to enhance dental appearance.

Further Reading: Jones, A. "Dental Transfigurements in Borneo." *British Dental Journal* 191 (2001): 98–102; Illich, Ivan. *Limits to Medicine: Medical Nemesis, the Expropriation of Health*. London: Marion Boyars, 1999; Johnson, Clarke. "Scarification, Mutilation, Dental and Body Alteration," http://www.uic.edu/classes/osci/osci590/13_4%20Scarification_%20Mutilation_%20Dental%20and%20Body%20Alteration.htm (accessed February 2008); Marenko, Betty. "Sink Your Teeth into This! Alternative Dentistry and Shiny Smiles ... "http://www.coldsteel.co.uk/articles/dentist.html (accessed February 2008); Ring, Malvin. *Dentistry: An Illustrated History*. New York: Harry N. Abrams, 1992; Rumsey, Nicola, and Harcourt Diane. *The Psychology of Appearance*. New York: McGraw-Hill, 2005; Smith, Bernard, and Howe, Leslie. *Planning and Making Crowns and Bridges*. 4th edition. Abingdon: Taylor & Francis, 2006; Tayanin G. L., and D. Brathall. "Black Teeth: Beauty or Caries Prevention? Practice and Beliefs of the Kammu People." *Community Dentistry Oral Epidemiology* 34 (2006): 81–86.

Peter G. Robinson

Teeth Filing

Teeth filing is a term broadly applicable to a number of practices that intentionally aim to change the shape of human teeth. Historically, such deliberate modification of the teeth has been documented to have been practiced to

Filed teeth, shown on this woman in Sumatra, Indonesia, in 2001, remain a beauty tradition for some. With pressure from the government, this tradition is slowly losing its importance in Indonesia. © Remi Benali/Corbis.

varying degrees in Africa, North America, Mesoamerica, South America, India, Southeast Asia, the Malay Archipelago, the Philippines, New Guinea, Japan, and Oceania. Due to the dilutive effects of the influx of Western culture, however, the practice is generally in decline, having already vanished from certain areas where it was once prevalent.

The term "filing" has not been universally used in anthropological, ethnographic, and archaeological literature, and as such intentional transformation of the teeth is also referred to as, among other terms, sharpening, chipping, shaping, modification, and culturally induced dental alteration depending on the temporal, geographic, and technical context of the modification in question. The classificatory system used to categorize types of tooth modification lists seven basic types. Each type has a minimum of five variants, and fifty-nine distinct modifications are identified in total. Deliberate filing changes the tooth by altering the dental crown (such as producing a sharp point) or its surface (by creating patterns or grooves on the front of the tooth), or via a combination of the two techniques. Each localized population employing these techniques sought a culturally specific aesthetic, using unique configurations of contour and surface changes. Certain populations in North America, South America, and in the Pacific also implanted precious metal inlays, jewels, or gemstones into holes bored into the teeth. For a large number of these cultures, it has proven difficult if not impossible to uncover the origins of such dental modifications, although speculations have been made which suggest aesthetic, initiatory, or magical intentions.

Changing the shape of the teeth is enormously painful, and many accounts describe that the procedures involved can quite dramatically shorten the lifespan of those who undergo it. Nevertheless, these factors have done little to dampen the historical popularity of tooth filing. Furthermore, even with relatively primitive surgical methods in comparison with the advanced dental technologies of today, evidence of damage to the tooth's structures which might cause short-term complications such as infections, cavities, or decay, is rare, and it is clear that most of these early dentists did have a certain amount of knowledge regarding oral anatomy.

The well-documented cases of dental modification come from the pre-Columbian civilizations of Mesoamerica, chiefly among the Mayans. Archaeological evidence suggests that the early Mayans (ca. 2000 BCE) practiced aesthetically simple forms of tooth modification, sharpening them to a point, but that the complexity of their decoration increased over time such that, by the end of the first millennium CE, the practice had become rather elaborate, particularly in the use of ornamentation. Several skulls uncovered in modern-day Mexico, for example, have teeth inlaid with jade and pyrite. These modifications generally occur on young adults of both sexes, and the relationship between the presence of dental modification and social rank or status is unclear. The Mesoamerican use of tooth filing appears to be principally decorative rather than functional, though as some religious icons have depicted saw-tooth deities, there may have been a religious or devotional aspect. Methodologically, it is probable that the simpler operations were carried out with stones, though the more complex procedures utilized special tools such as drills fashioned from bone or cane.

It has been established that tooth modification has also been practiced elsewhere in the Americas, with a great number of archaeological specimens discovered in Ecuador. The fashions there seemed to primarily favor precious inlays, although some filing has also been noted. In North America, deliberately mutilated or modified teeth have been found at various Native American sites in areas including Arizona, Tennessee, California, Georgia, and the Caribbean. Some specimens from the area of modern-day Texas are thought to date from the Archaic period (8000–1000 BCE), and these early samples, from both male and female jaws, bear very simple modifications in the form of a series of notches ground out of the tooth's biting edge. Remains from the largest prehistoric site in the United States, Cahokia Mounds near St. Louis, Illinois, show similar practices, though from a much later time period. The site, which is the major source of modified teeth samples in North America, encompasses a number of large cemeteries, and certain burial grounds reserved for lower-status individuals have yielded deliberately notched male and female incisors dated from between 1000 and 1400 CE. It is not known whether those of a higher-status within Cahokian society also engaged in similar notching.

In Panama, some young men and women from the Guyami tribe still transform their teeth into points, but it is not believed that this is a continuation of a practice from prehistoric times. Rather, it has been suggested that the fashion was appropriated from escaped African slaves in the seventeenth century, particularly as the chipping method employed did not resemble the abrasive technique used by ancient American civilizations. Tooth filing was historically practiced by a number of tribes across Africa (with the exception of the north), but it has become ever rarer since European colonization.

In Africa, tribes which modify the shape of their teeth do not tend to limit themselves to using files as the prehistoric Americans had. More commonly, traditional cosmetic dentistry in an African context chipped the teeth with knives, chisels, or similar implements, and the range of ornamentation found is subsequently rather different in comparison with the Americas. As in Mesoamerica, though, the purpose of these undertakings is unclear in the majority of cases: pointed teeth are simply entrenched aesthetic features of many tribal cultures, and when interviewed by ethnologists, even those with pointed teeth are unable to fully articulate the practice's background. In some cases, tribes ascribe an initiatory role; the Ibo, from Nigeria, used to prevent young women from having children until their teeth were pointed. Some early ethnographers dared to morbidly correlate sharpened teeth with cannibalism, but this is no longer considered a particularly credible hypothesis. More typical responses tend to accentuate a simple aesthetic justification, emphasizing either their effectiveness in making warriors appear more fearsome or the peculiar attractiveness or beauty of pointed teeth.

The Tiv, a pagan tribe from the Benue Valley in Nigeria, are an excellent example of the pursuit of vanity through teeth filing. Beauty is a hugely important concept in Tiv culture, both for men and women, and the chipping of teeth is just a part of a much larger beauty regime which also involves polishing the body with oils, wearing dramatic clothing, and incising the skin with complex scarification patterns. The combination of these modifications is intended to be visually spectacular and sexually attractive, with the particularly painful oral surgery being thought of very highly. Tooth modification is bluntly cosmetic, but the production of beauty through pain is in itself attractive: not only is it basely indicative of strength, virility, and tenacity of spirit, it also demonstrates a commitment to the production of visual pleasure for others and a focused pursuit of aesthetic refinement that is to be admired.

Once common throughout the Malay Archipelago, tooth filing (as well as other bodily rituals such as tattooing) was explicitly banned after Indonesian independence in 1954 and has become uncommon (though not totally unknown) as a result. Until then, it was customary among certain Indonesian tribes, such as the Iban in Borneo and the Batak from North Sumatra, to undergo teeth filing, as it was indicative of maturity and demarcated the passage into adulthood. They would rub leaves or a sticky tar-like substance onto their teeth to blacken them, file their upper incisors into sharp points with rough stones, and grind their lower teeth down to half their natural size. For young boys, this occurred at puberty, and for young girls, it could start as early as seven years of age. Girls must have had their teeth filed before their first menstruation, as it was believed that the filer might become contaminated otherwise. Once the teeth had been filed, a young member of these tribes was allowed to partake of the ingestion of betel, a mild narcotic that is an integral part of ritual and quotidian life throughout the archipelago. This ritual belief also engendered the equation of filed teeth with beauty, and although pearly white teeth were naturally prevalent, they were considered hideously unattractive.

Further Reading: Alt, Kurt W., and Sandra L. Pichler. "Artificial Modifications of Human Teeth." In Kurt W. Alt et al., eds. *Dental Anthropology: Fundamentals, Limits and Prospects*. Vienna: Springer-Verlag, 1998, pp. 387–415; Bohannan, Paul. "Beauty and

Scarification among the Tiv." *Man* 56 (September 1956): 117–121; Fastlicht, Samuel. *Tooth Mutilations and Dentistry in Pre-Columbian Mexico*. Chicago: Quintessence Books, 1976; Milner, George R., and Clark Spencer Larsen. "Teeth as Artifacts of Human Behavior: Intentional Mutilation and Accidental Modification." In Mark A. Kelley and Clark Spencer Larsen, eds. *Advances in Dental Anthropology*. New York: Wiley-Liss, 1991, pp. 357–378 Saville, Marshall H. "Pre-Columbian Decoration of the Teeth in Ecuador. With Some Account of the Occurrence of the Custom in Other Parts of North and South America." *American Anthropologist (New Series)* 15/3 (1913): 377–394.

Matthew Lodder

Testicles

Castration

Castration is a multivalent term, which in the broadest sense can refer to almost any amputation of genital tissue, but which normally refers to the removal of testicles in men. Here the latter definition is used, although several other important usages are mentioned. The traditional term for men who have been castrated is "eunuch," and it remains in use among scholars, particularly when discussing history. Today, castrated men do not refer to themselves as eunuchs, as the term has become pejorative.

Medicine and Sexuality

In contemporary Western medicine, surgical castration (orchiectomy and orchidectomy) is performed as a treatment for testicular and advanced prostate cancer. For treating prostate cancer, the procedure was first described in 1941. It reduces testosterone levels by 95 percent, and is remarkably effective at stopping cancerous growth. Today, such surgery can be done as an outpatient procedure with local anesthesia. An incision is made along the midline of the scrotum (where the scar will blend with the natural line of the perineal raphe) or above the penis shaft (where the scar will be concealed by hair). The surrounding connective tissues are cut to expose the testicles one at a time, the spermatic cord is tied to prevent bleeding, the testicles are severed and extracted, and the wound is sewn closed. Should a patient wish, prosthetic testicles can be inserted into the scrotum. Such prostheses make some men feel whole again, or please patients for their ability to create the illusion of intact genitals. The first prostheses were used in 1941, and were made from the alloy vitallium. Glass, Lucite, Gelfoam, polyethylene, polyester, saline-filled bags, solid silicone rubber, and silicone gel in a silicone rubber shell also have been used. Implants containing silicone gel are the most lifelike, but are controversial because the material causes health risks for women when used as breast implants. (Medical data on ruptured testicular prostheses is too scant to assess safety—as of 2002, only five cases had been reported, often caused by pressure through bike racing, horse jumping, or fetishistic sex.)

The surgical procedure has been declining in frequency, as castration by drugs that suppress hormones have become more popular. This suppression can be accomplished with drugs that that block hormone reception, hormone production, or control the hypothalamus and pituitary glands. Taken together, surgical and drug therapies are referred to as androgen-deprivation therapy. More than 40,000 men in North America start such therapy each year. While drug therapy avoids the psychological difficulty of having one's "manhood" disappear, surgery is simpler, cheaper, and more convenient. A drawback of drug therapy is that patients often forget or refuse to take their medications. After either form, men experience menopausal-like symptoms, such as hot flashes, fatigue, nausea, indigestion, enlarged or tender breasts, weight gain, sweating, weakness, impotence, change of skin and hair texture, osteoporosis, anemia, depression, loss of libido, and glucose intolerance.

The sexuality of castrated men is poorly understood. Although castration usually reduces the libido, many men retain desire, particularly if castrated after puberty. Although estimates vary widely, upward of 25 percent retain a sex drive. They may not have erections with predictability, but can engage in a panoply of other sexual and sensual practices with their partners. In some societies, eunuchs have been thought of as a third sex because their bodies have secondary-sex characteristics of both men and women. Castration is also one of the surgical procedures performed on male-to-female transsexuals, and indeed castrated men in many cultures worldwide have functioned as women or a third gender. As such, castration may be sexually liberating, allowing eunuchs to have relationships with both genders and explore socially taboo pleasures.

A little studied subculture of men who want to be or have already been castrated has emerged in contemporary society, primarily on the Internet. The average age of participants within the subculture is sixty, and the majority of the men are college-educated. The reasons for seeking out castration are manifold, but about 40 percent desire a "eunuch calm," 30 percent are sexually excited by the idea, and another 30 percent find castration aesthetically pleasing. The culture centers around Web sites such as The Eunuch Archive (http://www.eunuch.org) and *Body Modification E-zine* (http://www.bmezine.org). Because of the difficulty of finding medical personnel to castrate a healthy man, the community often resorts to self-mutilation and dangerous "street cutters." Unfortunately, these procedures often go badly, resulting in the need for professional medical attention.

Castration has been an important concept in psychiatry since Sigmund Freud (1856–1939) proposed in 1925 that boys developed "castration anxiety" after seeing female genitals, intuitively believing that their penis was cut off. Freud incorrectly believed that girls do not experience vaginal pleasure until after puberty, so they also fixate on their lack of a penis, developing "penis envy." Although Freud's central idea about males has retained professional respect, psychiatrists Melanie Klein (1882–1960) and Ernest Jones (1879–1958) pointed out that his ideas reduced females' experiences to a mirror of males' and proposed a more nuanced understanding of the psychosexual development of girls.

Myth and Religion

Both the practice and condemnation of castration are featured in the world's major mythological and religious systems. In Greek and Roman mythology, Cybele, a goddess of fertility, had a male partner, Attis, who castrated himself. As early as 415 BCE this served as a foundation for a cult of priests called the Galli, who dedicated themselves to Cybele after performances that culminated in ritual castration. The cult was unwelcome in Greece, but was received well in Rome, where temples were located. Like Attis, the Galli wore female clothing. Although the symbolism is poorly understood, by castrating themselves the priests may have strengthened Attis for his resurrection or become more similar to Cybele, to whom they were devoted.

Within the Abrahamic traditions of Judaism, Christianity, and Islam, castration has been treated in numerous ways. Castration has always been prohibited within the Jewish faith. Passages from the scriptures refer to this, including Deuteronomy 23:1, Leviticus 21:20, and 22:24 (New Revised Standard Version). Specifically, no man "whose testicles are crushed," may "be admitted to the assembly of the Lord." Also, no animal with "its testicles ... crushed or torn or cut" could be sacrificed. Later interpretation in the Talmud extended this prohibition to include any purposeful impairment to the reproductive systems. At the same time, Isaiah prophesied that, "To the eunuchs who keep my sabbaths, who ... hold fast my covenant, [the Lord] will give ... a monument and a name better than sons and daughters ... "(56:3-5). There are medical exceptions made to the Jewish laws.

Historically, Christianity has been ambivalent about castration. The Roman Catholic Church at various occasions condemned self-castration, such as at the Council of Nicea in 325 CE and a decree by Pope Leo I in 395 CE. Nevertheless, Jesus is reported to have taught: "For there are eunuchs who have been so from birth, and there are eunuchs who have been made eunuchs by others, and there are eunuchs who have made themselves eunuchs for the sake of the kingdom of heaven. Let anyone accept this who can" (Matthew 19:12). In 203 CE, Origen (185–254 CE), one of the church fathers, reportedly castrated himself because of this passage, but later another father, Augustine (354–430 CE), forcefully condemned the practice. Augustine argued that Jesus allegorically meant that Christians should abstain from marriage. Due to the competitive religious culture at the time, Augustine and others may have been setting Christianity up against what they considered to be pagan religions, including the cult of Cybele. Also, a religion requiring mutilation would understandably attract fewer adherents—this may have been the reason why the apostle Paul taught against circumcision of new Christian converts, a practice found in Judaism. Interestingly, possibly the first gentile convert to Christianity may have been a eunuch; a court official from ancient Ethiopia was converted by Philip in the book of Acts (8:26-40).

The positive treatment of eunuchs within Christianity is demonstrated by their consistent welcome in membership, their ability to be ordained if involuntary castrated, and their importance in church music. The castrati were men who were castrated before puberty and prized for their voices as adults. Such early castration slows the development of the vocal cords and larynx, resulting in high voices, an unusual vocal tone quality, sometimes a wider vocal range, and a larger breathing capacity. As church music developed, castrati were necessary due to the prohibition of women's voices in church. Although the church condemned castration, it paradoxically justified the castrati as promoting the public welfare with good music. Castrati were used within Western churches beginning in the late sixteenth century and ending in the early twentieth; eunuchs were present within the eastern churches from the fifth century. Castrati were also featured in opera, primarily in Italy during the seventeenth and eighteenth centuries. Composers such as Handel (1685–1759), Mozart (1756–1791), and Rossini (1792–1868) composed music calling for castrati voices. One of the most famous of castrati was Carlo Broschi (1705–1782), known as Farinelli, on whom an award-winning 1994 movie by the same name is based. During their apex, castrati could command high fees, but the form of opera in which castrati parts were composed declined. The last operatic castrato was Giovanni Battista Velluti (1780–1861), who retired in 1830.

As definitively as it is in Judaism, castration is forbidden in Islam; Mohammad forbid both self-castration and castration of others. This did not prevent Muslim rulers, however, from hiring physicians of different religious traditions to perform the procedure on their subjects; importing eunuchs from Africa, India, and elsewhere; employing them as palace and harem guards and in the military; and sending them as gifts. Eunuchs have been documented within Islamic empires as early as the Abbasid caliphate (750–1258 CE). Eunuchs even guarded Muhammad's tomb in Medina, a practice that ended in the 1920s; these were the last eunuchs under Islamic rule.

In China, castration was practiced from at least the eighth century BCE to 1911 CE. Although religion in China is difficult to generalize because of the varying practices and hybridization of Daoism, Confucianism, and Buddhism, eunuchs were often equated with deformity. They were permitted to enter temples but denied access to the altars of main deities. Because sterile men do not have descendants, it was difficult to ensure living caretakers for their souls. A common strategy to prevent this problem was to become a monk, as the community would ensure proper postmortem ritual. Such rejection of castration was also present in Chinese secular society. Eunuchs were despised by many and insulted for their odor if they were incontinent.

The eunuchs themselves were allowed to enter the Forbidden City and often attained political power as civilians with direct access to the royal family. Within the harem quarters, they served as guards, companions, and musicians. Princes were raised by eunuchs. As such, castrated men had the opportunity to mold the character of future emperors. The most powerful eunuchs in China were castrated as boys—sold by their families for service in the palace—but grown men could volunteer to be castrated for employment. Those under age ten were permitted the most intimate access to the harem, including bathing. For adults, the privilege of castration required testimony of good character, and although employment tasks were menial, the pay was high.

In China, castration included the removal of the penile shaft as well as the testicles, and unlike most parts of the world, documentation exists on the procedure. An account by the nineteenth-century Englishman George Carter

Stent explained that the patient was given an anesthetic drink and the genitals were numbed with pepper. The testicles and shaft were severed with a curved knife, the urethra plugged with metal, and paper bandages applied over the wound. After three days, during which the patient could not drink, the bandages were removed to allow urination. Full recovery took up to three months, and the procedure proved fatal for 2 percent of patients. The penis was saved in a jar of alcohol and inspected before professional promotions—if lost, a replacement could be purchased or borrowed. Upon death, a eunuch was buried with his penis. The last Chinese royal eunuch, Sun Yaoting, died in 1996—eighty-five years after the fall of the imperial government.

Although not technically castration because it refers to the severance of the penile shaft, the "vagina dentata" or "toothed vagina"—an orifice that bites off any penis that penetrates it—is a common myth worldwide. Forms of the myth appear in Native American, Indian, Maori, Hawaii, Greenland, Greek, and Christian cultures. In recent years, technologies used by women to simulate the vagina dentata have appeared in popular culture, such as the science fiction novel *Snow Crash* from 1992 by Neal Stephenson, actualized in 2005 as the "Rapex"—a latex tube worn like a tampon, filled with needle-like barbs that would require surgical removal. The vagina dentata was of particular interest among avant-garde artists such as Salvador Dalí (1904–1989) and Pablo Picasso (1881–1973)—sometimes shown as the female praying mantis who eats the male after copulating. Although completely unfounded, according to lore among American troops during the Vietnam War, local women inserted razor blades, chards of glass, sand, or other items into their vaginas as an insurgency technique, in effect realizing the myth.

Punishment

Castration has been used as a form of punishment for centuries, usually for sexual crimes or as a means to debase one's enemies. It was punishment for adultery in ancient Egypt, for rape in twelfth-century Western Europe, and for homosexuality in thirteenth-century France. Some evidence suggests prisoners of war were castrated in Europe during the Middle Ages. Even today, men reportedly have been castrated and left to die in the Sudanese Darfur conflict.

Castration as a form of punishment is rare or prohibited in developed countries. In the United States, the Supreme Court decided in 1910 in the case *Weems v. United States* that castration was "barbaric," although, in at least one state, California, a person convicted of child molestation can be castrated either chemically or physically. Sex offenders in the Federal Republic of Germany during the 1970s could choose castration in exchange for sentence reduction. A famous study (1989) by Reinhard Wille and Klaus M. Beier showed that 3 percent of those who chose surgical castration offended again, versus a rate of 46 percent for those who had chosen castration but later changed their minds. Sex offenders in some U.S. states are now chemically castrated, meaning that they are given medication to reduce the sex drive as a way to prevent recidivism. These drugs can be effective, but the number of sex offenders with abnormal sexual drives and fantasies that would logically merit such treatment is small, totaling some 10 percent of sex offenders. The American Psychiatric Association supports this chemical castration, but only after a clinical evaluation. Some contend the treatment may violate the Eighth Amendment to the U.S. Constitution, since courts in the past have prohibited medicating prisoners against their will. The medications have side effects—the same as those described above for cancer treatment. Finally, some psychologists say the drugs do not address the emotional problems that lead to the sexual offenses. *See also* Penis: Penis Envy; and Testicles: Eunuchs in Antiquity.

For Further Reading: Anderson, Mary M. *Hidden Power: The Palace Eunuchs of Imperial China*. Buffalo, NY: Prometheus, 1990; Aucoin, Michael William, and Richard Joel Wassersug. "The Sexuality and Social Performance of Androgen-Deprived (Castrated) Men Throughout History: Implications for Modern-Day Cancer Patients." *Social Science and Medicine* 63 (2006): 3,162–3,173; Bailey, J. Michael, and Aaron S. Greenberg. "The Science and Ethics of Castration: Lessons from the Morse Case." *Northwestern University Law Review* Summer (1998): 1,225–1,245; Bukowski, Timothy P. "Testicular Prostheses." In *Urologic Prosthesis: The Complete, Practical Guide to Devices, Their Implementation and Patient Follow Up*, Culley C. Carson, ed., 141–154. Totowa, NJ: Humana Press, 2002; Faison, Seth. "The Death of the Last Emperor's Last Eunuch." *New York Times*, December 20, 1996; Freud, Sigmund. "Some Psychological Consequences of the Anatomical Distinction between the Sexes." In *The Standard Edition of the Complete Psychological Works of Sigmund Freud*, edited by James Strachey and in collaboration with Anna Freud, 243–258. London: Hogarth Press, 1925, 1953; Friedman, David M. *A Mind of Its Own: A Cultural History of the Penis*. New York: Free Press, 2001; Gulzow, Monte, and Carol Mitchell. "'Vagina Dentata' and 'Incurable Venereal Disease' Legends from the Viet Nam War." *Western Folklore* 39, no. 4 (1980): 306–316; Hutson, Thomas E. "Management of Newly Diagnosed Metastatic Disease." In *Current Clinical Urology: Management of Prostate Cancer*, Eric A. Klein, ed., 561–578. Totowa, NJ: Humana, 2004; Jackson, Bruce. "Vagina Dentata and Cystic Teratoma." *The Journal of American Folklore* 84, no. 333 (1971): 341–342; Kuefler, Mathew S. "Castration and Eunuchism in the Middle Ages." In *Handbook of Medieval Sexuality*, Vern L Bullough and James A. Brundage, eds., 279–306. New York: Garland, 1996; Markus, Ruth. "Surrealism's Praying Mantis and Castrating Woman." *Woman's Art Journal* 21, no. 1 (2000): 33–39; Marmon, Shaun E. *Eunuchs and Sacred Boundaries in Islamic Society*. New York: Oxford University Press, 1995; Miller, Robert D. "Chemical Castration of Sex Offenders: Treatment or Punishment?" In *Protecting Society from Sexually Dangerous Offenders: Law, Justice, and Therapy*, Bruce J. Winick and John Q. La Fond, eds., 249–263. Washington, D.C.: American Psychological Association, 2003; Raitt, Jill. "The 'Vagina Dentata' and the 'Immaculatus Uterus Divini Fontis.'" *Journal of the American Academy of Religion* 48, no. 3 (1980): 415–31; Rosselli, John. "The Castrati as a Professional Group and a Social Phenomenon." *Acta Musicologica* 60, fasc. 2 (May–August 1988): 143–179; Taylor, Gary. *Castration: An Abbreviated History of Western Manhood*. New York: Routledge, 2000; Wassersug, Richard J., Sari A. Zelenietz, and G. Farrell Squire. "New Age Eunuchs: Motivation and Rationale for Voluntary Castration." *Archives of Sexual Behavior* 33, no. 5 (2004): 433–442; Whitby, Bob. "My Life as a Eunuch." *SF Weekly*, June 28, 2000, http://www.sfweekly.com/2000-06-28/news/my-life-as-a-eunuch; Wille, Reinhard, and Klaus M. Beier. "Castration in Germany." *Annals of Sex Research* 2 (1989): 105–109.

Travis Nygard and Alec Sonsteby

Eunuchs in Antiquity

"Eunuch" is a term that has been used since Greco-Roman antiquity to describe partially and fully castrated males. In classical antiquity, partially castrated eunuchs had their testicles damaged or removed before or after the onset of puberty, while fully castrated males also had their penises removed. The legendary Assyrian queen Semiramis, a mythical figure sometimes associated with the Babylonian queen Shammuramat (ruled 811–808 BCE), was reported in antiquity to have been the first to castrate males, when she sent 500 castrati to Persia to serve as catamites (boys used in pederasty). However, references to eunuchs appear as early as the Assyrian Law Code of Hammurabi (1810–1750 BCE), which has three legal provisions concerning adoption of the children of eunuchs. Indigenous to the east, castrated males were brought into Greece and Rome largely as a consequence of the slave trade. Eunuchs, who were typically slaves or manumitted (freed) slaves, served a variety of secular and religious functions and frequently rose to positions of considerable social and political influence.

Ancient Near East and Mediterranean

The ancient Greek word *eunouchos* means "guarding the bed," and referred in classical Greek literature to castrated males in the ruling courts of the Persian empire, who served as guardians of harems of the Persian king and other ruling elites. Largely because they could not have offspring who might inspire dynastic aspirations, eunuchs were often selected to serve in other, more elevated social and political positions, including as chief counselors, administrators, and military officers. One eunuch named Bagoas, for example, served as the chief commander for Persia's army under Artaxerxes III. Artoxares, a eunuch from Paphlagonia, became one of the most influential advisors to Darius II until he was executed in 424 BCE for having staged a coup against the king. Another Bagoas, Darius III's favorite catamite, was presented as a gift to Alexander the Great, and reportedly exerted a great deal of influence over the Macedonian king.

Castration occurred largely in consequence of kidnapping, pirating, trafficking in persons, and war, and may also have occurred in the ancient Near East as a form of legal punishment for sexual offenders. In his history of Egypt, Diodorus of Sicily, who wrote during the first century BCE, claimed that castration was one of the forms of punishment for male adulterers and for men who raped women of free, elite status. The historian Herodotus, writing in the fifth century BCE, mentions that the Persian king, Darius I, demanded that the Babylonians he had recently conquered send annually to the Persian court 1,000 talents of silver and 500 Babylonian boys to be castrated and to serve as eunuchs. And, when the Persians destroyed the Greek cities of Asia Minor in the events leading up to the Persian War, the conquering generals selected the most beautiful girls and boys from among the prisoners of war, castrated the boys, and sent them as presents to the Persian king. The castration of the cities' male youths not only supplied the court with eunuchs; it also symbolized for the conquerors the subjected, enslaved status of the Ionians living in Greek cities on the west coast of Asia Minor. Unusually beautiful boys were particularly esteemed in Persia, and could earn prodigious sums for the slave traders (some of whom were Greek) who castrated these kidnapped, pirated, or otherwise enslaved youths and sold them in the slave markets of Sardis or Ephesus.

Castration is mentioned in the earliest Greek literature, and appears as a fairly common theme in Greek myth. In Homer's *The Odyssey* (roughly eighth century BCE), Odysseus cuts off the ears and genitals of the traitorous suitor Melanthius, and throws them to the dogs to eat before killing his victim. And, in *The Iliad*, Priam appeals to Hector not to face Achilles on the battlefield by claiming that when Hector is dead, and the Greeks slay Priam, they will throw his dead body to the dogs, who will consume his genitals. In his account of the origins of the Greek gods, Hesiod's *Theogony* recounts the castration of the first supreme, male deity, Ouranos (Heaven) by his son Cronus. The goddess Aphrodite, according to Hesiod, was born from Cronus' castrated genitals that fell into the sea. Hesiod also mentions the *korybantes*, protectors of the infant Zeus. These nine armored, castrated priests of the goddess Rhea drowned out the cries of the infant Zeus with drumming, clashing of their weapons, and ecstatic dancing.

The classical Greeks, however, who probably rarely imported eunuchs into the mainland, expressed disdain for castration as a symbol of the corrupting luxury and effeminacy of the Persian elite. After allied Greeks successfully warded off the Persian invasions in the fifth century BCE, the eunuch in Greek art and literature served as a particularly effective icon by which to convey pejorative attitudes about Persian culture, which were often couched in terms of exotic barbarism and the corrupting influences of tyranny and excessive wealth. In the fifth-century Athenian comedy *The Acharnians*, for example, Aristophanes satirizes two Athenian politicians whom the protagonist recognizes as disguised (and traitorous) eunuchs attending the Persian ambassador Pseudartabus. While classical Greeks expressed contempt for the social practice of castration, nevertheless, ancient sources indicate that Greeks involved in the trafficking of humans, particularly in Asia Minor, were accessory to systematized castration because they supplied its victims for the niche, eunuch market.

Eunuchs in Roman Culture

Eunuchs in the Roman Empire served both sacral and profane functions. Archaeological and literary evidence attests to eunuch priests in a variety of ascetic and orgiastic cults throughout the ancient Mediterranean, including at least one cult of Hekate in Caria, Asia Minor, a cult of Artemis at Ephesus, the *korybantes* of Crete, and the priests and high priests of the Phrygian goddess Cybele.

Eunuchs first arrived in Rome as a consequence of the importation from Asia Minor of the cult of Attis and Cybele, whom the Romans called *Magna Mater*. Attis was a semi-deity who, in legend, was the exceedingly beautiful consort of Cybele. When Attis was sent by his parents to marry the daughter of a local king, Cybele, who had fallen in love with Attis, appeared to him in the form of an epiphany. The sight of the goddess drove Attis mad, and in a state of ecstatic frenzy, he cut off his genitals and dedicated himself to her service. An etiological story for the origin of the cult, the myth accounts for the orgiastic nature of its rituals and its priestly college of eunuchs.

The official adoption of the Magna Mater cult occurred, according to the Roman historian Livy, in 204 BCE, and this eventually brought about the arrival of priests from this Phrygian cult to Rome. Known as *Galli*, the priests of the Cybele and Attis were, according to ancient sources, males who castrated themselves in the fury of ecstatic worship. Serving under a high priest, the Galli were considered to be neither male nor female, and were variously called *eunuchi*, *spadones* (from a Greek word, meaning "to tear" or "rip"), *semiviri* and *semimares* (half-men), and *nec viri nec feminae* (neither men nor women) by Roman authors.

In his *Fasti*, the poet Ovid (43 BCE–17 CE) describes a festival called the *Ludi Megalenses* held on April 4 to honor the goddess Cybele, in which a parade of eunuchs marches while beating on hollow drums, clashing cymbals, and carrying an effigy of Cybele through the streets. A fourth century CE imperial calendar marks March 15–25 as festival days in honor of Cybele; these featured processions of flute players, the burial of an effigy of Attis, a day of mourning, a day of bloodletting and bull sacrifice, a day celebrating the resurrection of Attis and the consequent return of agricultural fertility, and a day of ritual purification. The popularity of this fertility cult grew and spread throughout the Roman Empire.

While the cult of Cybele flourished and generated a class of sacral eunuchs, trafficking in humans also contributed to the importation of eunuchs into Rome, particularly following the dissolution of the Roman Republic. As the elite classes grew wealthier through imperial domination, the sale and purchase of eunuch slaves played a part in the economics of conspicuous consumption. In wealthy households, eunuchs were synonymous with decadence, luxury, exoticism, and effete sensibilities. Castrati commanded some of the highest prices in the slave trade. While slaves intended to perform manual or domestic labor could be purchased for about 500 *denarii*, at least one source identifies a eunuch who was sold for 2,000 denarii. Pliny the Elder in the first century CE describes the purchase of a eunuch for the extraordinary price of 50 million *sesterces*.

It is likely that forced castration was performed on many youths, particularly those judged to have good looks. Victims of castration might be sold into prostitution. Others might be purchased by the elite to serve a variety of functions in the household or to be presented as gifts to patrons, clients, or peers. In such cases, eunuchs may have been forced to submit to the sexual whims of their masters. The Roman satirist Juvenal (late first to early second century CE) complained about women whose lasciviousness was so great that they had physicians remove the testicles of particularly handsome male slaves before they physically matured so that the women could avoid pregnancy and abortion.

By late antiquity, numerous sources offer anecdotes of eunuchs who exerted considerable influence on the emperor in matters of state and who served as powerful administrators of finance and as procurators. Tacitus reports that the Roman emperor Nero (37–68 CE) placed a eunuch of freed status, Anicetus, at the head of the Roman fleet. The fourth century CE Roman historian Ammianus Marcellinus complained that the eunuch Eusebius was running the government, and the emperor Justinian (482/3–565 CE) made the eunuch-chamberlain Narses commander of Roman troops in Italy.

Ancient historians often refer to eunuchs as examples of the excesses of the Roman emperors. Cassius Dio, for example, reports that the emperor Septimius Severus (146–211 CE) castrated 100 Roman citizens to serve as his daughter's attendants. One of the most infamous imperial eunuchs was a freedman called Sporus, who gained notoriety as a sexual partner of the emperor Nero. In his history of Rome, Cassius Dio claims that Nero so missed his deceased second wife, Poppaea Sabina, that he caused Sporus, a boy of freed-person status who resembled Sabina, to be castrated and to serve as Nero's catamite. Having supplied funds for Sporus' dowry, Nero is said to have held a public marriage ceremony for himself and Sporus, and to have taken the eunuch on a honeymoon to Greece.

Roman jurists commented sparingly on the legal status and rights of eunuchs. Castration appears to have been outlawed within the Roman Empire, and the Roman jurist Paulus said that the castration of any man against his will was deserving of capital punishment. The exact nature of castration, however, was up for debate. Discussing whether or not marriage between a *spado* (one of the terms for eunuch) and a woman was possible, Roman jurist Ulpian concludes that a juror should determine whether or not the eunuch's genitals had been mutilated by castration, or if he has not been castrated. A castrated male may not marry, since it is not possible to arrange a dowry, while a non-castrated male may marry, since it is possible to arrange a dowry. By a *non castratus spado* ("a eunuch who has not been castrated"), most scholars understand Ulpian to mean sterile or infertile (Frier and McGinn, 2004) or men whose sexual organs underdeveloped after puberty (Kuefler, 2001). Elsewhere, Ulpian distinguishes between a eunuch who is a spado by nature, one who has become a *thlibiae thlasiae*, which are derivates of the words meaning "pressing" and "crushing," and one who has become a spado by other means. *Thlibiae* referred to the practice of tightly tying up the scrotum to cut the vas deferens, while *thlasia* sterilized the testicles by crushing them.

Further Reading: Bruce W. Frier, and Thomas A. J. McGinn. *A Casebook on Roman Family Law.* Oxford: Oxford University Press, 2004; Kuefler, Matthew. *The Manly Eunuch: Masculinity, Gender Ambiguity, and Christian Theology in Late Antiquity.* Chicago: University of Chicago Press, 2001; Lane, Eugene, N., ed. *Cybele, Attis and Related Cults: Essays in Memory of M. J. Vermaseren.* Leiden: E. J. Brill, 1996; Scholz, Piotr O. *Eunuchs and Castrati: A Cultural History.* Trans. by John A. Broadwin. Princeton, NJ: Markus Wiener Publishers, 2001; Stevenson, Walter. "The Rise of Eunuchs in Greco-Roman Antiquity." *Journal of the History of Sexuality* 5:4 (1995): 494–511; Vermaseren, Maarten Jozef. *Cybele and Attis: The Myth and the Cult.* London: Thames & Hudson, 1977.

Angela Gosetti-Murrayjohn

Thigh

Liposuction of the Thigh

The female leg is a site that has incredible potential to erotically stimulate its onlooker. The meeting point of the female leg as the erotic locus renders the thigh an incredibly potent symbol of female sexuality. Cultural attention to legs has fueled Western women's obsession with the shape and size of their thighs, and has inspired much interest in cosmetic surgery.

The surgical removal of fat and skin from the thighs dates back to 1886, but the early surgeries were always accompanied by scars, which often became the locus for new fat deposits. It wasn't until the 1970s that a significant effort was successfully made to reduce and hide postoperative scars. Various methods of subcutaneous surgery (removal of fat from beneath the skin) using sharp curettes were experimented with; however, these procedures were inaccurate and carried significant risk of bleeding and nerve and blood vessel damage.

Italian gynecologist Giorgio Fischer invented a method in 1974 that is now called liposuction. He proposed to internally cut away at the loose connective tissue that stores fat, called adipose tissue, through small incisions. He used an electric rotating scalpel; once the fat was cut away, it was suctioned out with rubber tubing and a vacuum. This procedure was imprecise and resulted in blood loss, numbness, asymmetry, and other unpredictable results. Early liposuction left much room for improvement, but spawned a tremendous amount of methodological experimentation among cosmetic surgeons.

In 1977, French plastic surgeon Yves-Gerard Illouz developed what is known as modern liposuction, using blunt instruments to create tunnels to pass between major blood vessels. The instrument produced less bleeding and was easier to manipulate through fatty tissue and thus resulted in far less scarring. As with all forms of cosmetic surgery, the improvement of the procedure is measured partly by its increasing invisibility.

Tumescent liposculpture was developed in the United States in 1987. This procedure made possible the safer removal of large amounts of fat (approximately six to eight pounds per procedure), without general anesthesia of significant blood loss. It also surpassed old liposuction procedures in its ability to sculpt the area under scrutiny rather than simply to remove unwanted fat. Because the smaller cannulas that are used cause far less bleeding, the surgeon has more time to maneuver the desired shape.

As with all forms of cosmetic surgery, the discourse surrounding the liposuction (or liposculpture) of the thigh depends upon the prevalence of body dissatisfaction and raises the possibility of meeting a socially constructed ideal. Liposuction is often sold as an answer to the "problem" that there are certain parts of the body that, based on heredity, cannot be improved or diminished with diet and exercise. The thigh, divided in the plastic surgery field into sections (outer thigh, inner thigh, back of thighs, and front of thighs), is one of the areas of the body that is supposedly resistant to weight loss; the resultant figure "problems" are both stigmatized and medicalized, and the surgical solution is advertised as a mandatory last resort. Moreover, as with other forms of cosmetic surgery, liposuction is heralded as an answer to the physical effects of aging. For example, in *Body Contouring: The New Art of Liposculpture* (1997), the natural effects of aging are conflated with hereditary body disfigurement. This framework suggests that everyone is a candidate for liposculpture. The book uses drawings of the "ideal" female and male body to establish a norm against which the reader is expected to measure her or himself. The expectation projected by the authors is that everyone can and should liposculpture themselves into an idealized body type.

In many non-European cultures, the buttocks and legs surpass the breast as the primary symbolic site of female sexuality. The fetishization of the thighs, however, has extended to Western culture as well, and the natural aging process as well as the tendency for fat deposits to collect in this region drives people to the knife. Popular cultural references for the fat accumulation on the upper, outer thigh, "saddlebags," and popular disdain for the dimpling on the back of the thigh, or cellulite, code the upper leg as a problem area often in need of surgical intervention.

Further Reading: Coleman III, William P., C. William Hanke, William R. Cooke Jr., and Rhoda S. Narins. *Body Contouring: The New Art of Liposculpture.* Carmel, IN: I. L. Cooper, 1997; Gilman, Sander L. *Making the Body Beautiful: A Cultural History of Aesthetic Surgery.* Princeton, NJ: Princeton University Press, 1999; Morris, Desmond. *The Naked Woman: A Study of the Female Body.* New York: Thomas Dunne Books, 2004.

Alana Welch

Tongue

Tongue Splitting

Tongue splitting is the process of laterally bisecting the tongue. The tongue is essentially a number of symmetrically arranged muscles, and by cutting the fibrous tissue which runs down its center, it is possible to divide these muscle groups into two separate halves, creating a forked tongue. The procedure has emerged as one of the more common pseudosurgical modification procedures among the subcultural body-modification community since the turn of the twenty-first century. It has not achieved the cultural traction of activities such as piercing or tattooing.

Forked tongues have been historical staples of numerous folk and religious imageries for centuries, usually bound up in the rich symbolism of the snake, serpent, or dragon. In Buddhism, for example, the snake-like spirits known as naga have forked tongues, and in Hinduism, the serpent-god Kaliya also bears an anguine tongue. In the Christian tradition, the Roman poet Prudentius' work *Hamartigenia* even ascribes the origin of sin to a bifurcated tongue: on his fall, Satan's tongue becomes forked, and as the physical division both literally and metaphorically corrupts his speech, he is consequently able to lead humanity astray. Despite the long history and culturally broad significance of the image of the split tongue, there is nevertheless no real evidence that it had ever been practiced as an actual modification procedure until 1994 at the very earliest.

In June 1997, images and a brief e-mail were sent to the online body-modification archive http://www.bmezine.com by an anonymous Italian man from Florence. In his e-mail, the correspondent described how, three years earlier, a close friend of his had asked an acquaintance who was a dentist to slice his tongue in half. After applying a light local anesthetic, the dentist used a scalpel to divide his patient's tongue into two parts and then cauterized the resultant wounds with silver nitrate, producing the world's first documented tongue split.

At the beginning of 1997, before the Italian's experience appeared online, a twenty-year-old body piercer from California named Dustin Allor acquired the world's second split tongue. Apparently entirely independently, an idea had begun to germinate in Dustin's mind that she might be able to use her existing tongue piercings as the base for dividing her tongue. She threaded thin fishing line through one of her piercings, tied it into a knot around the tip of her tongue and pulled tight. With a great deal of persistence and a not-inconsiderable amount of pain and discomfort, by tightening the line each day, she was able to fully close then knot and thus split her tongue entirely in two after a period of about three weeks.

Dustin's split was published as the cover story of *Body Play and Modern Primitives Quarterly* at the end of 1997. In the meantime, unaware of Dustin's self-performed split, sideshow performer Erik Sprague approached oral surgeon Lawrence Busino in Albany, New York, clutching images of the enigmatic Florentine. After an apparently brief consultation, Busino agreed to perform the operation using an argon biopsy laser more usually used to file teeth cavities. Known professionally by his stage name "The Lizardman," Erik was already becoming a prominent figure in the body-modification community and beyond, and once news of his successful split spread, the popularity of the procedure began to snowball.

As more and more people became aware of tongue splitting, it soon came to the attention of the authorities. In 2001, State Rep. William Callahan from Michigan sponsored the world's first piece of legislation aiming to explicitly outlaw the splitting of the tongue. Michigan legislature's House Bill 4688, an amendment to the state's public health code, was narrowly defeated and did not pass into law. Nevertheless, the case engendered concern in legislators across the United States, and in 2003, Rep. David Miller successfully introduced a bill in Illinois that made the carrying out of a tongue split by an unlicensed individual a criminal code violation. Following the Illinois example, tongue splitting is now restricted or prohibited in numerous states, including Pennsylvania, Delaware, and Texas. The U.S. military also explicitly bans servicemen and women from having a split tongue at all.

Legislative attention has deterred medical professionals from performing tongue splits, and as a result, professionally executed splits using lasers and similar dental and surgical technologies are now particularly rare. Nevertheless, the negative attention of lawmakers seems to have done little to dampen the subcultural interest in tongue splitting, and it remains popular among those interested in pursuing body modifications beyond tattooing and piercing. In terms of technique, some modifiers still employ Dustin's tie-off approach, but it is generally considered to be an ordeal. Scalpelling has become the most commonly used method as it is relatively simple and quick to perform, and while simply cutting the tongue in two with a scalpel can be quite bloody, healing is relatively unproblematic. Some practitioners prefer to use sutures or cauterization on the inner surface of the wound as the two halves of the tongue can often fuse entirely back together if care is not taken to pull them apart during the first few weeks of healing, but this practice is far from universal.

Once healed and with a little practice, each half of the tongue can be moved independently of the other. There is no noticeable negative effect on either speech or taste. It is impossible to ascribe widespread reasons or ritualistic intentions to the practice, other than within the general narrative of ownership and corporeal exploration that pervades the subcultural rhetoric of the body-modification community.

Further Reading: Benecke, Mark. "First Report of Nonpsychotic Self-Cannibalism (Autophagy), Tongue Splitting, and Scar Patterns (Scarification) as an Extreme Form of Cultural Body Modification in a Western Civilization." *American Journal of Forensic Medical Pathology* 20/3 (September 1999): 281–285; Musafar, Fakir. "Dustin's Split Tongue." *Body Play and Modern Primitives Quarterly* 4/4 (1997): 14–20 Sprague, Erik. "The Modern History of Tongue Splitting," http://www.bmezine.com/news/lizardman/20050726.html (accessed April 2007).

Matthew Lodder

Uterus

Chastity Belt

The chastity belt is a locked device worn as an undergarment that is meant to create a physical barrier to sexual intercourse or masturbation. Images of medieval iron underwear or belts worn with a padlock are widespread, connoting harsh attempts to control women and their sexuality. The medieval chastity belt is largely mythical, but the chastity belt has remained an important symbol of female subordination. Recently, the chastity belt was revived in sexual fetish practices. The chastity belt is also the subject of some contemporary human rights controversies.

The chastity belt was first presented by Conrad Kyeser in his book *Bellifortis*, on the subject of military technology, in 1405; however, his belt was a conceptual sketch and no actual model from this era was ever found. Medieval poetry contains references to chastity belts made of cloth, which according to the poets were worn consensually as a pledge of faith. Art from the fifteenth and sixteenth centuries portrays women donning chastity belts over their genital regions, and even padlocks on their tongues. However, the literal cages and padlocks of the medieval period are largely myths created in the Victorian era, when fascination with the chastity belt was high.

Italian women of the Renaissance purportedly wore chastity belts. Recently, however, chastity belts were removed from museum displays in France, Germany, and England when their authenticity as Renaissance artifacts was questioned. Historians have suggested that some Victorian working women donned locked undergarments for protection, but primarily, chastity belt-like devices were promoted for the prevention of masturbation. From the eighteenth century to the 1930s, masturbation was widely regarded as harmful and deviant, and various devices were utilized in an effort to prevent it. In addition to the chastity belt, however, were other methods. Boys' pants were designed so that they couldn't touch their genitals through their pockets, girls were often banned from riding horses and bicycles, and bland diets were generally imposed as they were thought to dull one's sexual impulses.

While the chastity belt enjoyed only limited use, it can be understood as a potent symbol of sexual subordination. Feminists have long criticized

Seventeenth-century chastity belt on display at the Musee de Cluny in Paris. © Bettmann/CORBIS.

the treatment of women as the sexual property of men, and have noted the ways in which standards of virginity have been unevenly applied to the sexes. Demands for female chastity have been deeply embedded across cultures and persist in a variety of moral codes and laws in modern societies. In fact, the United States even today operates with a federal agenda concerned with the virginity of its unmarried citizens. The Adolescent Family Life Act, or as it was more colloquially referred, "The Chastity Act," was instituted as recently as 1982, its goal to spread the message that abstinence is the only valid sexual option for unmarried persons. More recently the Bush administration promoted abstinence-only education as the solution to teenage pregnancy. In the United States, unmarried parenthood has been framed as catastrophic and unacceptable. In contrast, high numbers of births to single women in Denmark, Sweden, and Iceland are met with considerably less moral panic.

Contemporary Chastity Belts

The contemporary version of the chastity belt appears almost exclusively for use in sexual fetish practices such as sadomasochism or bondage/discipline/sadomasochism (BDSM). Both a costume and a device used to achieve pleasure through sexual denial, the modern use of the chastity belt, in most cases, reflects personal sexual expression rather than the imposition of sexually conservative mores. Chastity belts used in BDSM play are designed for both men and women. Contemporary chastity belts sold as fetish gear are often made of leather and metal or plastic and have padlocks. Chastity belts have been featured regularly in pornography, and are sometimes referenced in contemporary lingerie. In 1970, the chastity belt was the inspiration for a Cartier-designed "love bracelet" that is attached with a screwdriver, making it difficult to remove. Some conservative Christian teenagers pledge sexual abstinence before marriage by wearing so-called "purity" rings or necklaces, which are supposed to be removed if the wearer engages in sexual intercourse.

While these examples are related to sexual consent and agency, some instances of the revival of the chastity belt raise significant human rights concerns. The Asian Human Rights Commission (AHRC) filed a report in 2007 that women in Rajasthan, a state in northwest India, were being forced to wear chastity belts. The AHRC reports that women in Rajasthan are widely abused. Women are considered by some local party leaders, mostly members of the Bharatiya Janata Party, to be equal to *dalit* (or lower caste), and are treated unequally in many sectors of society. Abduction and trafficking of young women and girls to profit from the sex industry are common, and violence against women is rarely prosecuted.

Reminiscent of reports of English working women of the nineteenth century who used chastity belts for protection, there are a few modern instances of chastity belts and similar devices being promoted to protect females from rape. In 2001, a newspaper article in the South African daily newspaper *Sowetan* featured an image of a chastity belt-like device for infant girls, suggested as a solution to the increasing problem of rape of infants. Also in South Africa, an anti-rape device that is supposed to be inserted into a woman's vagina was developed by Sonette Ehlers. The device would cut into a penis penetrating the vagina; it can be removed only with medical assistance. The proposal to sell it sparked outrage among women and ignited a public debate over the high number of rapes occurring in the country. A recent estimate (2002) by the South African Law Commission puts the number of rapes at almost 1.7 million annually. A recent controversy in Malaysia was sparked when a prominent cleric, Abu Hassan Al-Hafiz, suggested that women wear chastity belts to protect themselves from rape. Global women's activists and human rights campaigners have condemned all of these developments, likening them to medieval practices. The suggestion that women should "lock" their genitals to prevent rape raises significant concerns for the welfare and human rights of women and children.

Further Reading: Blank, Hanne. *Virgin: The Untouched History.* New York: Bloomsbury, 2007; Classen, Albrecht. *The Medieval Chastity Belt: A Myth-Making Process.* New York: Palgrave, 2007; Davis, Natalie Zemon, and Arlette Farge, eds. *A History of Women: Renaissance and Enlightenment Paradoxes.* Cambridge, MA: The Belknap Press of Harvard University Press, 1993; Klapisch-Zuber, Christiane, ed. *A History of Women: Silences of the Middle Ages.* Cambridge, MA: The Belknap Press of Harvard University Press, 1992; Nuttall, Sarah. "Girl Bodies." *Social Text* 22.1 (2004): 17–33.

Alana Welch and Victoria Pitts-Taylor

Cultural History of the Vagina

The vagina, or the muscular canal leading from a woman's external genitalia to the cervix of her uterus, is a site of sexual and reproductive importance as well as of cultural, social, and political interest and controversy. The vagina has often been seen to possess potency and power, inciting both veneration and scorn. A survey of medical, cultural, and religious representations of the vagina reveals a wide range of attitudes toward and treatment of the vagina. For example, in androcentric views in the history of Western medicine, female genitalia have been viewed as undeveloped or inferior versions of male organs, while belief systems in India and China have considered female genitalia as the symbolic origin of the world and the source of all new life.

Physiology

The vagina, meaning "sheath" or "scabbard" in Latin, is a muscular canal of about six to seven inches in length that leads from the cervix to the outside of the body. It is lined with mucus membrane, joining the cervical opening at one end to the labia minora at the other. During sexual intercourse, the penis enters the vagina (thus the term "sheath," which refers to the outer covering of a sword). Sperm can enter the cervical opening to reach the inner reproductive organs. The vagina is also called the birth canal because of its role during childbirth, providing passage for the fetus. The hymen, named for the Greek god of marriage, Hymen, is a membrane that partially covers the entrance to the vagina and is associated with virginity, since it is believed that an intact hymen can be broken during sexual intercourse.

Since antiquity, medical writers have largely described the female role in sexual reproduction as passive. The vagina has been perceived primarily as a receptacle for male "seed" and has been seen as largely inert. However, recent research regarding reproduction reveals evidence to contradict the perception of the vagina, and the woman to whom it belongs, as a passive vessel. Studies are showing the vagina to be a complex organ that has a significant role in the outcome of sexual reproduction. For example, the vagina plays a major role in sperm selection. The path that sperm must traverse in order to reach a female's egg is now being described as a dangerous obstacle course; sperm that make it through, based on the sophisticated system of female-mediated sperm transport, are likely to be the strongest and most genetically promising.

Views of the Vagina in Western Medicine and Culture

Medical writings on women's bodies date to antiquity. Accounts of female bodies make up nearly one-fifth of the Hippocratic *Corpus*, the medical writings attributed to Hippocrates from the fifth and fourth centuries BCE; distinctly gynecological texts from the fourth to the fifteenth centuries CE number more than 100. Until the nineteenth century, however, medical understandings of female bodies were largely informed by the view that women were incomplete or undeveloped men. Female genitalia were considered interior forms of male genitalia. The two-sex model in the nineteenth century overturned this view, describing two distinct sexes, with distinctly male or female genital organs.

The vagina has been referred to in various translations of the *Trotula*, a compendium of medical writings collected in the twelfth century CE, as "genitalia membra" (genital membranes), as a "weket" (gate), "loca secreta" (secret place), or "pudibunda" (shameful parts). The *Trotula*, written by several authors, including probably an Italian female physician, described various treatments for illnesses and conditions of the vagina and for injuries from childbirth, including ointments, herbs, soaks, powders, pessaries, and tampons. The *Trotula* also offers advice for women who may not have an intact hymen. The text describes the use of leeches to create blood in the vagina on the night before a woman is married, for the purpose of producing proof of virginity. Later medical texts, including one from the fifteenth century, list other methods, such as inserting into the vagina substances from oxen that mimic blood, and inserting matter that makes the vaginal walls bleed.

The Hymen

The hymen itself has a controversial place in the history of medicine. The very existence of the hymen has been debated by medical writers for centuries; according to historical debates, it is either a myth, a membrane that is broken during intercourse (or before), or a part of the vagina itself. The first discussion of the hymen as a specific membrane covering part of the opening of the vagina appears in the second century CE in a text by Soranus: he describes the belief in its existence as "erroneous." Nonetheless, the hymen was infused with deep significance. In the Renaissance, a medical exam of the hymen could constitute proof of virginity. (For centuries earlier, bloodstained sheets on the wedding night served as evidence of a bride's virginity). Midwives were also sometimes responsible for inspecting the hymen for signs of virginity or sexual activity.

The intact hymen was so symbolically laden in the sixteenth century that having sex with a virgin was thought to cure a man of syphilis. Similar myths are currently seen in relation to AIDS, and have been associated with a high incidence of rape of girls in some countries. In South Africa in 2000, more than 21,000 cases of reported rapes had children or infants as their victims, according to statistics compiled by South Africa's police service.

While the medical relationship to the hymen has overwhelmingly been aimed at preserving, inspecting, or providing evidence of virginity, in the mid-twentieth century, some American doctors believed that an intact hymen was an impediment to marital harmony, because it created a physical and psychological barrier to marital intercourse. The fear of pain from breaking the hymen, they theorized, could cause sexual discord in a newly married couple, and recommended either the breaking of the hymen in a nonsexual, medical setting (hymenotomy) or exercises to dilate the hymen prior to marriage.

Hysteria

Although medieval physicians were concerned with virginity, classical physicians appeared to worry even more about sexual abstinence. The condition of hysteria and its accompanying ailments was believed to be caused by female sexual abstinence. Ancient Greek physicians, including Hippocrates (fourth century BCE) and Galen (second century CE), believed that sexual abstinence was unhealthy, and that in women it caused a range of problems, including the "wandering womb." The diagnosis of hysteria was based on classical humoral theory, which argued that health depended upon the balance of humors, or fluids (yellow and black bile, blood, and phlegm) in the body. Sexual intercourse was believed to regulate the flow of menses, while abstinence was believed to cause a uterine buildup of humors. In addition, sexual intercourse, along with pregnancy, was perceived to help anchor the womb, because a dry womb would gravitate toward moister organs. Marriage and intercourse were recommended to treat hysteria, although early physicians did not recommend sex outside of marriage. For the unmarried, relief could be obtained through massage stimulation by a physician or midwife that culminated in orgasm, or the "hysterical paroxysm." (Ironically, in the nineteenth century, clitoridectomy was also proposed as a solution to female disorders, primarily masturbation but also hysteria.) Hysteria was perceived as a significant medical problem through the Victorian era, eventually losing credibility as a medical diagnosis in the twentieth century.

The Vibrator

The orgasm was seen as a cure for hysteria, and its regular occurrence as precautionary and healthy. Since the time of Hippocrates, physicians had been recommending induced orgasm for their patients with hysteria, and often left the time-consuming job to midwives. The history of medical massage technologies culminates in the development of the vibrator. The vibrator was developed to literally free up the hands of doctors. The first electromechanical vibrator was invented in the 1880s; a steam-powered version was invented by an American doctor in 1872. Vibrators were sold to the public for home use since around 1900; in 1918, a vibrator appeared in the Sears and Roebuck catalogue as a home appliance. These products were advertised as home-based health remedies rather than as erotic toys. In fact, it wasn't until the vibrator appeared in early erotic American films in the 1920s that perceptions of the female vibrator became more pornographic than medical.

Vaginal Orgasm

While the medically induced orgasm faded from use, medical and psychological debates over the female orgasm intensified in the twentieth century. The Venetian anatomist Renaldus (or Realdo) Columbo (1516–1559) claimed to have discovered the clitoris as the "seat of women's pleasure" in 1559, but the early twentieth century saw a fierce debate over the physiology of the female orgasm. Sigmund Freud (1856–1939) argued in *Three Essays on the Theory of Sexuality* (1905) that there were two kinds of orgasms, vaginal and clitoral, with the former representing the mature organism of the grown woman. The vaginal orgasm, occurring when the vagina involuntarily contracts, was important for Freud because he saw it as the heterosexual and reproductive organism necessary, ultimately, for civilization. In Freudian terms, the clitoral orgasm was relegated to what were considered the developmental, aggressive, and unfeminine phases of early female sexuality.

However, in 1953 in his text *Sexual Behavior in the Human Female*, the famous sexologist Alfred Kinsey (1894–1956) located the female orgasm for most women solely in the clitoris. The 1966 text *Human Sexual Response* by sex therapists William Masters (1915–2001) and Virginia Johnson (1925–) rejected the division between two kinds of orgasm altogether, and argued that physiologically speaking, there is only one kind: a vaginal orgasm results when the clitoris experiences stimulation during intercourse. The feminist movement of the 1960s and 1970s embraced the acknowledgement of the clitoris and its role in female sexuality. In these debates, the clitoral orgasm became emblematic of women's independent or liberated sexual pleasures, partly because clitoral stimulation did not depend upon intercourse. The vaginal orgasm, on the other hand, became associated with a long history of medical and scientific androcentrism.

Cultural Representations and Practices of the Vagina

The Venerated Vagina

The traditionally disparaging view of the vagina and female sexuality in the West should not obscure the historical instances in which the vagina was venerated as an important symbol of fertility and strength. Spanning millennia and milieu, the custom of a woman raising her skirt to expose her genitalia has been an integral part of religious and cultural ritual. Dating back to ancient civilizations, the famous Greek explorer Herodotus, traveling through Egypt in the fifth century BCE, noted that gender roles there seemed an inverse of those in Greece. This female dominated culture had a religious custom in which female genital display, or as he named it, "anasuronai" (from the Greek word "to raise one's clothes"), was a defining part. This act has been interpreted cross-culturally as both evil-averting and fertility-enhancing. The vagina as a symbol of female power has been represented extensively in cultural practices. For example, *anlu* was a genital-exposing dance ritual practiced collectively against perpetrators of female abuse in West Cameroon. In medieval Europe, Sheela-na-Gig sculptures peppered Christian churches with their fertility-provoking vaginal display, an astonishing number of which survived their subsequent censure by the Christian church. Documented by anthropologists in the twentieth century, the Polynesian communities of the Marquesas Islands believed in the power of female genitalia to fend off evil supernatural spirits.

Female Genital Cutting

The vagina is not always venerated. Sometimes it is seen, along with the clitoris and the labia, as needing surgical transformation to be socially acceptable. One of the most controversial cultural practices regarding female genitalia is female genital cutting, also called female genital mutilation and female circumcision. Female genital cutting is characterized by a range of procedures, including cutting or removing the labia, the inner and outer lips surrounding the vaginal opening, removing the clitoral hood (prepuce) or the clitoris itself, or sewing together the outer vaginal lips. One type, the clitoridectomy, the removal of part or all of the clitoris, was proposed in European and American medicine as a treatment for a range of women's health problems, including hysteria and masturbation, in the nineteenth century. The last known procedure to be performed on medical grounds was in 1927. Clitoridectomy is no longer considered a legitimate medical practice in the West. Currently, however, the practice of female genital cutting for cultural reasons is widespread and far-reaching. It is prevalent in twenty-eight countries, mainly in North Africa and in some Middle Eastern and Asian regions, and is practiced by various diasporic communities in Europe, North America, Australia, and New Zealand.

The World Health Organization distinguishes between four types of female genital mutilation (FGM), their preferred term for female genital cutting. Type I consists of the removal of the clitoral prepuce (hood), with or without excision of all or part of the clitoris. Type II, the most common type, consists of total excision of the clitoris and partial or total removal of the labia minora (inner vaginal lips). Type III, also known as infibulation, is believed to have been first practiced by the early Egyptians. Infibulation involves the total removal of all external female genitalia, including the clitoris, prepuce, labia majora (outer vaginal lips), and labia minora (inner vaginal lips). The vaginal opening is then closed with stitches, leaving only a slight opening for the passage of menstrual blood and urine. Type IV can include cauterization or scraping of the vagina or external genitals, or the introduction of toxic substances into the vagina to constrict the vaginal canal.

Sometimes the procedure is performed by men, and sometimes by medical doctors, such as in Egypt, but usually a specially trained woman does the procedure. In some cultural contexts, the cutting is performed individually, while in others, groups of girls are cut in a ritualized ceremony. Women and girls are cut for a variety of reasons. The procedure may be seen as a significant rite of passage, and in some contexts, it is perceived as necessary to make a woman or girl eligible for marriage. Some believe that the elimination of the possibility of female orgasm or the closing of the vaginal opening is the only way to ensure chastity and morality. Some believe that sex with an uncut woman will not be pleasurable for men, or that the male member will be bothered by the clitoris (reminiscent of the mythical *vagina dentata*, or the "vagina with teeth").

The prevalence of female genital cutting has created a global controversy. The procedure is usually performed without anesthesia. The health risks can be significant, leading to infection, pain, a lifelong inability to enjoy sexual intercourse, and even death. Further, many critics, including the United Nations, describe the practice as a violation of basic human rights. Some would like to see a global ban of the practice, while others suggest that solutions must develop indigenously in the setting of their occurrence.

The Designer Vagina

In the debates over female genital cutting, some indigenous writers have argued that Western critics of female genital cutting should look to their own cultures for examples of mutilation. Surgical modifications of the vagina and labia are increasingly popular cosmetic surgery practices in the West. The so-called "designer vagina" refers to a surgically modified vagina that is designed to seem, presumably from a male partner's perspective, tight and virginal. Vaginal surgeries are also advertised as erasing the effects of childbirth on the vaginal canal. Vaginal-tightening surgeries, also called vaginoplasty or vaginal-rejuvenation surgery, originally developed out of a series of surgeries used to repair injuries from childbirth and to cure urinary incontinence. These procedures are now marketed to healthy women as ways to increase sexual pleasure during intercourse.

Also described as "sexual-enhancement surgery," cosmetic surgery of the vagina developed in the United States and remains controversial. The ordinary effects of age and childbirth are the targets of cosmetic surgeons performing vaginoplasty, and critics worry that cosmetic surgeons are pathologizing women's aging bodies, and labeling women's post-childbirth vaginas as abnormal. Feminists have pointed out that the procedures, marketed as able to enhance sexual experience, reinforce Freudian notions of vaginal orgasm and ignore the role of the clitoris in female pleasure. Relatedly, they worry that the surgery may not actually increase female pleasure, and is really aimed primarily at increasing pleasure of their male partners.

Pelvic Exercises

The Western history of body modification to tighten the vagina predates vaginoplasty. The gynecologist Arnold Kegel (1894–1981) developed a set of exercises to tighten and tone the vagina after childbirth, to treat incontinence, and to increase sexual pleasure. Kegel, like other physicians and psychiatrists of his time, believed that female orgasm was primarily vaginal, and that a proper orgasm occurred when the muscles of the vaginal canal contracted involuntarily during intercourse. He developed a program, known as Kegel exercises or pelvic-floor exercises, of voluntary muscular contractions of the pubococcygeus muscles, which form part of the pelvic floor and surround the vagina. Repeated exercises can treat vaginal prolapse.

Kegel exercises are preceded by a range of ancient practices aimed at tightening the vagina and increasing sexual pleasure. The Taoists of ancient China developed similar exercises, as did followers of the practice of Hatha yoga, a form of yoga developed in the fifteenth century in India. The *Kama Sutra*, a Sanskrit text written by Mallanaga Vatsyayana between the fourth and sixth centuries CE, describes an array of poses that are meant to increase sexual pleasure.

Reclaiming the Vagina

While an emphasis on female sexual pleasure seems like a recent phenomenon, the female orgasm is actually an age-old topic of interest.

Aristotle in the fourth century BC theorized that although women could conceive without orgasm, its occurrence greatly increased the chances of pregnancy. This perception of the female orgasm has been crucial to its moral acceptance by even the church and science. It is important to note, however, that acknowledgment, acceptance, and attention to the female orgasm has rested on an emphasis of its reproductive function, and a corresponding de-emphasis of its sexual and pleasure-giving functions. Women's pleasure for its own sake is now being addressed by a wide array of feminists and scholars.

For example, founded in New York City in 2000 by Melinda Gallagher and Emily Scarlet Kramer, Cake is a women's sexuality enterprise that arose as a way for young women in New York City to discuss their sexuality. Cake's vision for the next wave of feminism consists of demands for sexual empowerment and the freedom to explore and enjoy sexuality to attain gender equality. Efforts such as this one are proliferating throughout Western culture, raising the bar on female sexual expectation and fulfillment and promoting pride in and attention to one's genitalia. *See also* Genitals: Female Genital Cutting; Uterus: "The Wandering Womb" and Hysteria; Vagina: The Vagina in Childbirth; Vagina: *The Vagina Monologues*; and Vagina: Vaginal-Tightening Surgery.

Further Reading: Blackledge, Catherine. *The Story of V: A Natural History of Female Sexuality.* New Brunswick, NJ: Rutgers University Press, 2003; Drenth, Jelto. *The Origin of the World: Science and Fiction of the Vagina.* London: Reaktion Books, 2005; Gallagher, Melinda, and Emily Scarlet Kramer. *The Hot Woman's Handbook: The Cake Guide to Female Sexual Pleasure.* New York: Atria Books, 2005; Green, Monica Helen. "From 'Diseases of Women' to 'Secrets of Women': The Transformation of Gynecological Literature in the Later Middle Ages." *Journal of Medieval and Early Modern Studies*, vol. 30, no. 1 (Winter 2000): 5–39; Maines, Rachel P. *The Technology of Orgasm: "Hysteria," the Vibrator, and Women's Sexual Satisfaction.* Baltimore, MD: The Johns Hopkins University Press, 1999; Schwartz, Kit. *The Female Member: Being a Compendium of Facts, Figures, Foibles, and Anecdotes about the Loving Organ.* New York: St. Martin's Press, 1988; Sevely, Josephine Lowndes. *Eve's Secrets: A New Theory of Female Sexuality.* New York: Random House, 1987; Cake, http://www2.cakenyc.com.

Victoria Pitts-Taylor and Alana Welch

The "Wandering Womb" and Hysteria

The womb, *hystera* in Greek, is another term for the uterus, the pear-shaped organ in the lower abdomen of females that has a primary role in menstruation and pregnancy. The term "womb" is particularly associated with pregnancy, and in popular usage refers to the uterus' role as a protective, enclosed space for a developing embryo or fetus. The "wandering womb," a concept from Greco-Roman antiquity, not only describes supposed health problems associated with the womb, but also reflects the highly symbolic significance of female reproductive organs.

Ancient Greek medical writers purported that women's health was predicated on sexual intercourse with men, because coitus was thought to moisten the womb. According to these Hippocratic writers, the womb had a dry nature, and when the dry and empty womb was not anchored down by pregnancy or by the dampness produced during sexual intercourse, it had the potential to "wander," or drift upward and compress the intestines, liver, diaphragm, heart, and lungs. The symptoms produced by a wandering womb included shortness of breath, drowsiness, loss of speech, and madness (the medical term for which was "hysteria"). In extreme cases, it could even cause sudden death. Irregular menses could produce similar symptoms.

The Hippocratic corpus, a collection of more than sixty medical treatises composed between the fifth and fourth centuries BCE by several anonymous sources, gets its name from the writings attributed in antiquity to the physician, Hippocrates of Cos (fourth century BCE), often called the "father of medicine." While female patients and the illnesses of women are mentioned throughout the corpus, eleven of the treatises deal specifically with gynecological anatomy, disease, and treatment. Other important sources for ancient gynecological theories and treatments include: the philosophers Plato and Aristotle; the first century CE Roman medical writer Celsus; Aretaeus of Cappadocia (second century CE), who was greatly influenced by the Hippocratic corpus; Galen of Pergamum (second century CE), whose career began as a doctor of gladiators, but who eventually served as physician to Emperor Marcus Aurelius; and Soranus, another Cappadocian who practiced medicine and wrote medical treatises during the second century CE.

Symptoms

Although the Hippocratic writings are not entirely consistent, in general, they purport that the womb is the cause of all "women's diseases," because it has a tendency to become misshapen by prolapsis—when the uterus shifts due to lack of uterine support—and to migrate from the lower abdomen to other parts of the body. Intercourse and pregnancy reduced the risk of displaced wombs for two reasons. First, intercourse and pregnancy regulated the flow of menses, which was thought to produce numerous health problems if the blood did not flow freely. Early Greek medical writers thought of blood as one of the four humors, or substances which were believed to ebb and flow through the body. Good health demanded the correct proportional balance of the humors. Second, the womb had a tendency to gravitate toward moister organs or to become misshapen when too empty or dry. Sexual intercourse and pregnancy helped to maintain the womb's regular shape and to prevent the womb from bending over on itself, which could restrict the flow of blood out of the womb through the vaginal canal. Intercourse and pregnancy also opened up the pathway for the flow of blood during menses, because they stretched the walls of the womb and widened the opening of the cervix, encouraging the flow of blood.

Such theorics reinforced the importance of intimate heterosexual relations. In a treatise entitled *On Virgins*, one Hippocratic author claimed that virgins who are not married by the time they reach puberty are much more likely to experience hallucinations. Whereas a healthy woman's blood will flow like the blood of a "sacrificial animal," the restricted blood in the virgin wombs will backflow into the heart and lungs when it cannot flow regularly out of the vaginal pathway. When excess blood flows into the heart and lungs, physical and mental illness result. The medical writer stated that when these organs are filled with blood, so-called "wandering" fevers and chills are induced, which in turn lead to madness caused by excessive inflammation. Virgin girls can become fearful, murderous, and suicidal. They will try to choke themselves because of the pressure on their hearts, and in some cases, visions resulting from hallucinations prompt them to throw themselves into wells and drown. The writer claims that when virgins suffer this disorder, they should cohabit with men as quickly as possible.

Several ancient authors described the womb as an animal in their discussions of the symptoms and treatments of womb-related illnesses. Plato claimed that the animal-like womb will wander about the body when it is ready for conception, and in its wandering, will block air passages and cause suffocation. In his treatise, *On the Causes and Symptoms of Acute Diseases*, Aretaeus of Cappadocia likened the womb to an animal inside an animal. Like all animals, he claimed, the womb has a wandering nature, and it responds to certain smells, either being attracted to them or repelled by them. The symptoms that he described as occurring when the womb strays upward and compresses other organs are reminiscent of those in the Hippocratic corpus: the woman will experience choking, loss of respiration and speech, and epileptic-like symptoms, although without convulsions. Similarly, Celsus claimed that illness of the womb can produce symptoms which mimic epilepsy, causing a woman to suddenly fall to the ground without breath and to experience stupor; the symptoms of the wandering womb, however, differ from those of an epileptic seizure in that there is no rolling of the eyes or foaming at the mouth.

Treatment

Both the Hippocratic and Aretaeus' prescriptions for treatment of the wandering womb recommended a regimen of pessaries, or ointments applied to wool and inserted into the vagina. Sweet-smelling ointments will lure an upward-turned or vagrant womb down into proper position. Acrid or pungent-smelling ointments will cause a downward-turning or forward-turning, prolapsed womb to retreat upward into place. A Hippocratic author recommended for a patient who suffered from dropsy of the womb that she be given a laxative, a vapor bath made of cow dung, and a pessary made with ground-up cantharid beetle, followed several days later by a pessary of bile, and several days after that by a douche of vinegar. Soranus describes other drastic methods used by his predecessors, including the beating of metal plates and the playing of flutes (based on the presumption that the womb is also responsive to harsh or sweet sounds); applying constricting bandages around the midriff; blowing vinegar into the patient's nose (to force the womb downward into place); fumigating the patient's vagina with burning hair, wool, cedar resin, or dead bugs (to force the womb upward if prolapsed); and blowing air into the vagina using a blacksmith's bellow (to unblock the passage of blood so that menses could flow unimpeded). Suggesting that illness of the womb was caused by inflammation and not as a result of a uterus moving about "like a wild animal from the lair," Soranus rejected the wandering-womb theory.

The Female Physical Nature

Gynecological theories supported the ideology and practice of patriarchy, because they provided anatomical evidence that women were physically and mentally weaker than men and, therefore, more prone to fluctuations of emotion and sensibility. Aristotle, for example, discussing the role of women in reproduction, claimed that men and women differ in their *logos*—that is, in their capacity to reason. He suggested that this difference can be attributed to the fact that men have the capacity to generate in another (that is, through the deposition of semen in the womb), while women's generative power is self-contained (that is, women cannot impregnate others). Aristotle goes on to say that that which is female can be defined by a lack of power, vis-a-vis an inability to generate reproductive seed through the process of concoction. Departing from the dominant view of the Hippocratics, Aristotle argued that women do not contribute any seed in the process of reproduction. (The Hippocratics believed that the woman contributed a weaker and less perfect seed than the male, and that when weaker semen overwhelmed stronger semen during the process of conception, girls were the result.) Aristotle claimed that the womb functions solely as a vessel or receptacle for the male seed, which alone has generative power. Because semen is produced by a process of concoction that requires heat, and because women cannot produce semen, all males, according to Aristotle, are hotter by nature than females. The relative coldness of women not only prevents them from concocting reproductive seed, it also accounts for the abundance of blood that discharges from the body during menstruation.

The Hippocratics, however, believed that women's flesh had a porous physical constitution. According to the Hippocratics, while the male of every species is warmer and drier and his flesh dense and compact, the female is colder and wetter, and her flesh is sponge-like, soft, and moisture-retaining. Likewise, the natural state of the womb was moist, damp, and cool. When it became otherwise, physical and mental ailments resulted.

This account of the physical differences between men and women had far-reaching consequences, because, like Aristotle's theory arguing that women had a colder nature than men, the Hippocratic gynecological theories accounted for normative social beliefs that attributed to women physical, mental, and emotional inferiority. Because women were colder, according to these theories, they were more prone to sedentary lifestyles, they performed less labor, and they required fewer nutrients and food. Because women were more sedentary, they did not work off excess heat or moisture. And, because they were porous, women absorbed and retained moisture in the form of blood, and an accumulation of blood in the organs caused erratic and irrational behavior. In a healthy woman, this excess of blood purged every month in the form of menses, but, according to the

Hippocratics, porosity and sanguinity defined the female physiology. Therefore, the symptoms produced by excess blood—vicissitudes of temperament and irrationality—were also those that defined the female.

Medical understandings of female anatomy reflected power relations between men and women in antiquity; women's anatomy served as an explanation for a social hierarchy that invested men with legal, political, and familial authority over women. An Athenian woman in the fifth century BCE had to have a *kyrios*, or legally appointed male guardian, conduct any civic business on her behalf, including arranging marriages, pressing or defending against legal suites, and pursuing intestate and inheritance claims against potential rivals. The kyrios would also have to approve any medical treatment for gynecological or other maladies experienced by a woman. The Hippocratic prescription to "cohabit with a man as quickly as possible" to relieve hysterical suffocation brought on by irregular menses also reflects contemporary social practices: girls in ancient Greece often were married as early as twelve or thirteen to men typically in their thirties, and their most important duty to the household was to produce legitimate heirs. Throughout classical antiquity, girls often were married quite young. Soranus noted that the age for conception begins at fifteen in girls who are not too masculine-looking, muscular, or flabby.

Even as late as the second century CE, gynecological theories that attributed to women a weaker physical composition were still circulating. Galen of Pergamum, for example, claimed that the female is less perfect than the male because she has inferior reproductive capacities. He claimed that her reproductive organs formed within the body because a lack of sufficient heat prevented them from growing on the outside of the body. Galen also adhered to the Hippocratic tradition, inasmuch as he attributed to women the ability to generate reproductive semen, and claimed that the semen produced by women is colder, wetter, less vigorous, and less numerous than male semen because the physical nature of women is wet and cold.

Pregnancy Termination and Prevention

The medical treatises also offered advice on effective pregnancy termination and prevention. Pessaries and suppositories of various kinds were used for pregnancy prevention. Pennyroyal and other herbs were widely used as abortifacients. A Hippocratic author describes a case in which he advised a pregnant slave, employed by her mistress as a sex worker, to leap up and down seven times, touching her buttocks with her heels each time, since the slave's owner wished to terminate the woman's pregnancy in order to prevent loss of income. A loud thud followed the violent leaping, and the embryo fell to the ground. Other medical writers also recommended violent motion, together with applications of various kinds. Soranus recommended that women who wished to induce abortion should engage in the following: heavy exercise; be shaken by draught animals; use diuretics; eat pungent foods' inject douches of olive oil; take baths of very hot water mixed with fenugreek, marshmallow, wormwood, and other ingredients; and use poultices with oil, rue, absinthium, ox bile, and other substances. Arguing that it is better to prevent conception than to abort, Soranus also discussed a variety of contraceptive substances. He recommended that a woman smear the vaginal opening with honey, cedar, myrtle, balsam oil, or white lead; he also suggested that pine bark be rubbed with wine, wrapped in wool, inserted into the vaginal canal before intercourse, and extracted after. He further suggested that a variety of vaginal suppositories be used, such as a concoction of ground pomegranate peel, oak galls, and ginger, which were to be molded with wine into the size of a pea, dried, and inserted into the vaginal canal before coitus.

Hysteria in the Nineteenth and Twentieth Centuries

Ancient concern for the wandering womb formed the precedent for medical interest in various women's problems associated with the uterus and reproductive organs. In the nineteenth century, the term hysteria came into wide usage, linked to an array of physical and psychological symptoms, exclusively those of women. Hysteria was used as a diagnosis for numerous conditions, including unhappiness and anxiety. Because sexual frustration was thought to be a cause of hysteria, treatments included pelvic massage, or stimulation to orgasm by a midwife or physician. The diagnosis fell out of favor in the twentieth century, partly because it was so broadly used that it seemed to medicalize any female complaint. Today, the term "conversion disorder" is preferred, used more narrowly by psychiatrists to describe neurological symptoms that appear to have no physiological basis. *See also* Blood: Cultural History of Blood; and Vagina: Cultural History of the Vagina.

Further Reading: Fantham, Elaine, Helene Peet Foley, et al. *Women in the Classical World*. Oxford: Oxford University Press, 1994; Hanson, Ann Ellis. "The Medical Writers' Woman." In Froma Zeitlin et al., ed., *Before Sexuality: The Construction of Erotic Experience in the Ancient World*. Princeton, NJ: Princeton University Press, 1990, pp. 309–338; King, Helen. *Hippocrates' Woman: Reading the Female Body in Ancient Greece*. New York: Routledge, 1998; Lefkowitz, Mary R., and Maureen B. Fant. *Women's Life in Greece and Rome: A Source book in translation*. 2nd ed. Baltimore, MD: The Johns Hopkins University Press, 1992; Longrigg, James. *Greek Medicine From the Heroic to the Hellenistic Age: A Sourcebook*. New York: Routledge, 1998; Nutton, Vivian. *Ancient Medicine*. New York: Routledge, 2004.

Angela Gosetti-Murrayjohn

Vagina

The Vagina in Childbirth

The vagina is a muscular canal that extends for six to seven inches from its opening to the cervix, through which sperm enter to reach the uterus and ovaries. Its role in human sexuality is suggested by the original meaning of the word "vagina." Vagina is Latin for sheath or a scabbard, which is the container that holds a sword, a reference to the penis. This etymology might be said to reflect a male, heterosexual perspective that defines the female body in the terms of its male sexual counterpart. However, the vagina's role as the birth canal, or the route for childbirth, also influences its cultural meanings.

Until the seventeenth century, the vagina was not only the passage for the child, but was also considered to be in direct connection with the maternal mouth. One method to test a woman's fertility was to put a clove of garlic in her vagina, and if she later smelled like garlic out of the mouth, this was seen as a sign of fertility. During pregnancy and birth, the fetus was thought to breathe through the maternal mouth and the labia. In cases where a mother was thought to be dying, it was therefore advised to put a piece of wood between the her lips during birth and to spread her labia to allow the unborn child to breathe through this imagined channel until a cesarean section, a surgery that dates to antiquity, could be performed.

Historical sources describe a wide range of methods applied to the treatment of the vagina during vaginal birth. These are primarily aimed to ease the passage of the child through the birth canal. Greek and Roman authors in antiquity advised that the midwife massage the vagina with oils and other lubricants such as butter. In times when there was no water or bathrooms inside houses, midwives placed the woman on birth chairs with pots of steaming water underneath; the steam softened the vaginal tissue and eased her pain. In order to protect the perineum, the area between the vagina and the rectum, during birth, the main task of the midwife was to decrease the speed of the birth of the head in order to leave time for the vagina and the perineum to stretch and avoid tearing. Different ways to place the midwife's hands and how to position women for birth are described in old teaching books. Warm cloth was applied to the perineum to ease pain and enable further stretching. There were no techniques in many places for the suturing of tears; therefore, they needed to be avoided at all costs.

Once surgical repair techniques were developed, this changed. Episiotomy is a surgical cut into the perineum to widen the birth outlet. At the beginning of the twentieth century, episiotomy was reserved for exceptional cases, such as difficult breech births, abnormal size of the fetal head, or malformations of the maternal pelvis. In the second half of the twentieth century, however, a shift toward the routine use of episiotomy accompanied the medicalization of birth. Early twentieth-century medical textbooks argued that an episiotomy improves sexual function and decreases risks of postpartum fecal and urinary incontinence. In addition, with the change of the birth position from upright to lying down, from on the side to lithotomy (on the back, legs in stirrups), active pushing was introduced. A short delivery was considered best; one reason for an episiotomy was to decrease the time for the expulsion of the baby's head. This paradigm of episiotomy as a preventive method led to the increase of episiotomy rates in Western countries; they rose up to 90 percent in some places for women having their first baby. The rationale for episiotomy at subsequent deliveries was that scarred tissue cannot stretch; thus, many women had episiotomies at each birth.

With the increasing influence of evidence-based medicine and the publication of meta-analyses of studies examining the outcome of episiotomy, the routine use of episiotomy was challenged, but practice did not change much, and rates have dropped slowly. The Cochrane Library, an important source of evidence-based research, advises avoiding episiotomies except for rare situations in which the baby must be removed immediately. Rather than preventing tearing, episiotomies may increase the rate of tears and damage the perineum. There remains a significant difference in episiotomy rates according to the health care provider. The lowest episiotomy rates are found in midwifery-led care, whether in hospital or out-of-clinic settings. The highest rates are found in private obstetrical practices.

Although rates are decreasing in some countries, enormous variations in episiotomy rates between countries, especially between countries and even within caregiver groups. For countries, the rates vary between 9.7 percent in Sweden and nearly 100 percent in Taiwan. Small studies from Dublin and Denmark also suggest a wide variation of episiotomy rates among midwives. While some see strong policies and care guidelines as a method to decrease rates, others interpret the ongoing use of episiotomy as a cultural problem. One theory is that fear of litigation, and a profound lack of trust in the ability of the body, are the main reasons for the ongoing use of episiotomies. But critics suggest that while the episiotomy is conducted as a preventative measure, in practice, it creates the damage it seeks to avoid.

Episiotomies, the lithotomy position (flat on back with legs elevated), the rise in the use of cesarean sections, and other aspects of medicalized childbirth have been challenged in recent years by contemporary midwives and other critics of medicalization. The most severe criticisms have proposed that the routine use of episiotomy be considered a form of genital mutilation in Western obstetrics. While episiotomy rates are dropping only slowly, traditional methods are increasingly finding their way into mainstream care and are now being integrated into studies. With the increasing return of home birth and midwifery since the 1970s, old practices are being revived. Many midwives recommend massage of the perineum during the descent of the head, and it is now becoming common practice to warm the perineum during the birth of the head. In addition to massaging the perineum, an upright birth position, slow pushing, and slow delivery of the head is another recommendation. The birth position has been found to be the main influence on the status of the perineum, tears, and episiotomies. This is supported by historical sources that taught midwives to prevent extensive pushing by birthing women. New studies recommend avoiding this kind of pushing, and letting the woman push according to her urge, thus allowing a slow widening of the vaginal tissue.

In addition to these practices, the Kegel exercise, named after gynecologist Arnold Kegel (1894–1981), was developed in the mid-twentieth century as a method to train the pelvic floor muscles to increase the probability of an intact perineum; women were told to prepare for childbirth through dedicating themselves to regular practice of the Kegel. The Kegel, which remains widely used today, is not the first practice designed to tone the vagina; precedents for the Kegel include those developed by Indian yogis and those dating to ancient China. Another, recently developed method is the preparation of the vagina with a mechanical device called a vaginal obturator. In studies sponsored by the manufacturer, it has been shown to have a slight effect on the rate of episiotomies. Critics of such devices have suggested that they assume that the female body is unable to give birth naturally without technological preparation. Special preparations and technologies are needed.

Another aspect of the vagina in childbirth is the connection to sexuality. While in medicalized and particularly hospital birth, such a link is denied, some natural-birth centers and home-birth midwives promote the so-called "sensual birth," citing women reporting "cosmic orgasm" during birth. The sensual birth includes nipple and clitoral massage, which reportedly lead to decreasing pain, increasing relaxation, and the widening of the vagina. It has been argued that birth can be seen to be a sexual process because the same hormones are released as are during sex and lactation. For this release, a relaxed, trustful atmosphere and privacy are necessary. The point has also been made that the sexual context of birth has been taboo since male physicians took over the birth process.

In contrast to the notion of a sensual birth, many of the references to the vagina during birth in contemporary popular literature are connected to either disgust or pain. An example of such a description of the birth canal can be found in Eve Ensler's play *The Vagina Monologues*. While Ensler's description of the vagina is otherwise quite positive, the vagina in birth is presented as bleeding, tearing, and secreting slime and feces. One character in the play, a Ukrainian midwife, is described as digging around in the vagina, and the vagina itself is presented as a dwelling in which the child is stuck and waits to be freed. This is finally done with the help of modern medicine when the child is rescued via forceps. In Ensler's depiction, the birth canal turns into something mysterious and dangerous. It is a place the child better move through fast, a dangerous passage described with words similar to near-death experiences.

In psychoanalytic theory, birth represents the initial psychological trauma of one's life; the journey from the comfort of the womb to the outside world is one that generates great anxiety. The work of Freudian psychoanalyst Otto Rank (1884–1939), *The Trauma of Birth* (1929), was pivotal in developing this perspective, while Sigmund Freud (1856–1939) himself preferred the Oedipal view. Such a view of the vagina as a dangerous place of passage is also reflected in some modern branches of psychology, especially those focusing on perinatal psychology, developed in the 1970s by male psychologists and therapists such as Stanislav Grof (1931–) and Leonard Orr (1938–). Orr developed a technique of "rebirthing," in which patients were guided through breathing techniques that were meant to connect them with their birth experiences.

The attachment of a pure sexual meaning to the vagina on one hand and negative assumptions about the vagina as a birth canal on the other hand play a role in the increasing use of elective cesarean sections. Some women have used cesarean sections to entirely avoid vaginal stretching. Those who have not are made aware that the vagina that has undergone childbirth is now a site of surgical repair: elective vaginal-tightening surgery is an increasingly popular practice in cosmetic surgery clinics. A new aesthetic of the vagina is being generated: in the age of high-tech, elective medicine, the postchildbirth vagina suggests that the cosmetic surgery industry ought to feel and function as though childbirth was never experienced. *See also* Vagina: *The Vagina Monologues*.

Further Reading: Abers L., and N. Borders. "Minimising Genital Tract Trauma and Related Pain Following Spontaneous Vaginal Birth." *Journal of Midwifery & Women's Health* 52: 3 (2007): 246–253; Bumm E. *Grundriss zum Studium der Geburtshilfe.* Trans. by Katarina Rost. München und Wiesbaden: Verlag Bergmann, 1922; Ensler, Eve. *The Vagina Monologues*. New York: Random House, 1998; Davis E. *Heart and Hands. A Midwife's Guide to Pregnancy and Birth*. Berkeley, CA: Celestial Arts, 1987; Gaskin, Ina May. *Ina May's Guide to Childbirth*. New York: Bantam, 2003; Graham, Ian D. *Episiotomy: Challenging Obstetric Interventions*. Oxford: Blackwell Science, 1997; Gupta, J. K., G. J. Hofmeyr, and R. Smyth. "Position in the Second Stage of Labour for Women Without Epidural Anaesthesia." *Cochrane Database of Systematic Reviews* (2007); Hillebrenner, J., S. Wagenpfeil, R. Schuchardt, M. Schelling, K. F. Schneider. Trans. by Katarina Rost. "Initial Experiences with Primiparous Women using a New Kind of EPI-NO Labor Trainer." *Zeitschrift für Geburtshilfe und Neonatologie*. 205 (2007): 9–12; Kitzinger, S., J. M. Green, M. Keirse, K. Lindstrom, and E. Hemminki. "Roundtable Discussion Part 1." *Birth* 33 (2006): 154–159; Kok, J., K. H. Tan, P. S. Cheng, W. Y. Lim, M. L. Yew, and G. Hyeo. "Antenatal use of a Novel Vaginal Birth Training Device by Term Primiparous Women in Singapore." *Singapore Medical Journal*, vol. 45:7 (2004): 318–323; Marchisio, S., K. Ferraccioli, A. Barbieri, A. Porcelli, and M. Panella. "Care Pathways in Obstetrics: the Effectiveness in Reducing the Incidence of Episiotomy in Childbirth." *Journal of Nursing Management* 14 (2006): 538–543; Northrup, C. *Frauenkörper, Frauenweisheit*. Trans. by Katarina Rost. München: Verlag Zabert, 2001; Odent, Michael. *Geburt und Stillen. Über die Natur elementarer Erfahrungen*. Trans. by Katarina Rost. München: Beck C. H., 2000; Spitzer, B. *Der zweite Rosengarten. Eine Geschichte der*

Geburt. Trans. by Katarina Rost. Hannover: Elwin Staude Verlag, 1998; Stephens, A. "Oh Baby!; Forget Epidurals." *The Independent,* March 20, 2007, 12.

Katarina Rost

The Vagina Monologues

The Vagina Monologues, a play written by Eve Ensler that premiered off-Broadway in 1996, is based on her interviews with a diverse group of 200 women about their vaginas and their experiences with sexuality. This play is representative of the late-twentieth-century feminist concern that women's relationships with their bodies, and specifically with sexuality, have been repressed in a patriarchal culture. The play was initially performed by Ensler as a one-woman show at the Cornelia Street Cafe in New York City. Eventually, it was recast with an ever-changing trio of female actors and played at the off-Broadway Westside Theatre, at Madison Square Garden, and at other celebrated venues, and a televised version aired in the United States in 2001. The script consists of poems, facts, and monologues that constitute the storytelling of women's experiences with their vaginas, including accounts of rape and molestation, childhood discoveries of the vagina, orgasms, and memories of masturbation. The play is remarkable for its explicit language about female genitalia and about broader issues related to

Eve Ensler, right, creator of the off-Broadway show, *The Vagina Monologues,* is joined by, from left, feminist Gloria Steinem and actress Glenn Close at an after-party in New York for a "V-Day" benefit performance of Ensler's show, 2001. (AP Photo/Darla Khazei)

female sexuality. In 1997, the play won an Obie award, and the book has been translated into twenty-four languages.

The play also is credited with raising awareness of violence against women. In 1997, it became the inspiration for the founding of V-Day, a grassroots organization whose goal is to stop violence against women. Volunteers and college students put on fundraising performances of *The Vagina Monologues* on Valentine's Day, February 14, the proceeds from which go to efforts to fight violence against women. V-Day is global, and now includes organizations and performances across the United States and in Mexico, Egypt, Israel, Palestine, Jordan, Lebanon, Afghanistan, and many more countries. The organization's goal is to build anti-violence networks and to provide funds to grassroots, national, and international organizations that combat violence against women. Since its inception, V-Day has raised more than $30 million.

The Vagina Monologues has also ignited considerable controversy. The language and content of the play, including the very use of the word "vagina," has irritated conservative critics. Organizations such as the Cardinal Newman Society, a group dedicated to the revival of Catholic identity in Catholic higher education, and the American Society for the Defense of Tradition, Family and Property have been offended by the play's "explicit" and "vulgar" sexual content and have urged students and parents to protest it. In some cases, the play has been banned from certain venues as a result of such objections. For example, in March 2007, three female high school students from Cross River, New York, were suspended for their inclusion of the word "vagina" in their performance of scenes from *The Vagina Monologues*. While the principal insisted that their suspension resulted from their breach of an agreement to exclude the word from their performance rather than their use of the word itself, the girls, many parents in the community, and Ensler herself saw the punishment as blatant censorship. The play has also attracted criticism from pro-sex feminists who see it as overly focused on women's negative sexual experiences.

Ensler's other plays include *Lemonade*, *The Depot*, and *The Good Body*, which explores women's relationships with bodily practices such as cosmetic surgery, body piercing and tattooing, dieting, and exercise regimens.

Further Reading: "About V-Day," http://www.vday.org/contents/vday; Ensler, Eve. *The Vagina Monologues*. New York: Villard, 1998; Fitzgerald, Jim. "Girls Suspended Over 'Vagina Monologues.'" *San Francisco Chronicle*, http://www.sfgate.com/cgi-bin/article.cgi?f=/n/a/2007/03/06/entertainment/e134236S28.DTL&type=entertainment; "*The Vagina Monologues*: About the Book," http://www.randomhouse.com/features/ensler/vm/book.html

Alana Welch

Vaginal-Tightening Surgery

Vaginal tightening (sometimes referred to as vaginal rejuvenation or vaginoplasty), is a surgical procedure designed to decrease the diameter of the vagina, thereby countering the loss of the optimum structural architecture of the vagina associated with childbirth and/or aging. This, and other associated "designer vagina" procedures, developed out of earlier gynecological procedures to cure urinary stress incontinence and repair vaginal and perineum tears. Despite this, vaginal rejuvenation as it is currently conceived and marketed is framed primarily as a way to enhance sexual pleasure rather than as a cure for incontinence. It is sold to willing patients as a cutting-edge, liberating, elective procedure rather than as a mundane, routine, corrective one. While vaginal rejuvenation is relatively expensive and is not covered under health insurance in the United States or under the National Health Service in the United Kingdom, most clinics that provide this and other related procedures offer financing.

One surgeon at the forefront of the development of procedures intended to produce "designer vaginas" is David Matlock, a Los Angeles-based practitioner who has trademarked the procedures he has developed, namely, laser vaginal rejuvenation, and designer laser vaginoplasty. Such procedures, Matlock claims, produce ideal female genitalia, that is, genitalia that are "postvirginal and prechildbearing." This ideal state, and the sexual satisfaction presumed to accompany it, is constructed in the rhetoric surrounding such procedures as something to which women are "en*tight*led." In interviews, magazine articles, Web sites, and self-authored books, proponents of this particular form of "sexual-enhancement surgery" unapologetically appropriate the language and goals of liberal feminism, arguing that, historically, women's needs, desires and pleasures have been ignored, and that the time has come to recognize that women want and deserve sexual independence, pleasure, and satisfaction as much as men do. Such accounts, then, construct more and better "female sexual pleasure" as an unquestioned common good, as something that all rational, liberated women must surely pursue. This view of the enlightened, sexually autonomous, female consumer is accompanied by an image of the plastic surgeon as the profeminist savior of women who, until recently, have had to live with loose vaginas and less-than-pleasurable intercourse since this, it was commonly supposed, is the price to pay for motherhood.

A number of feminist criticisms and concerns have been raised in relation to vaginal rejuvenation and its social, medical, and ethical status in a market economy in which bodies (and selves) are increasingly commodified. One such concern regards the highly contentious way in which female sexual pleasure is conceived and constructed in material promoting so-called sexual-enhancement surgeries. Interestingly, while much of this material asserts that the procedures concerned are designed to empower women to become independent sexual subjects, vaginal tightness is nevertheless consistently represented as imperative, much more so than, for example, small, neat labia, or a suitably exposed clitoris. The message is that sexual pleasure is directly related to the amount of friction created between a penis and a vagina—that a tight vagina is the right vagina. On the other hand, a loose vagina—one that provides too little friction for the penis it is presumed to have been designed to envelop—will, according to Matlock, provide less erotic stimulation than, as he puts it, a crossword puzzle. If this is correct, and heterosexual men get little or no pleasure out of having to rotate their hips in circles during coitus because simple thrusting does not bring the penis into contact with the vaginal walls, then vaginal rejuvenation might be said to serve the beneficent social function of keeping a man from leaving for a younger woman. This problematic conflation of tightness with satisfaction clearly raises the question of whose pleasure is really at stake. In contradistinction to the second-wave feminist reclamation of the clitoris as the seat of female power and pleasure, the conception of female sexuality informing sexual-enhancement surgeries (and their promotion) remains firmly entrenched in the Freudian model which associates sexual maturation with a shift from the clitoris (as the primary source of pleasure) to the vagina, from autoeroticism to reproductive heterosexuality. In other words, while claiming to be a revolutionary and liberating practice for women, vaginal rejuvenation tends to reproduce heteronormative ideals about gender, sexuality, the body, and pleasure.

A related concern regards the medicalization and pathologization of what some might argue are inevitable physiological changes associated with aging and/or childbirth. In much of the material about "lax vagina syndrome" or "sexual dysfunction," vaginal loosening is represented as unhealthy, unfeminine, immodest, undesirable, and most definitely unsexy; it is associated with a loss of youthfulness, with sexual dissatisfaction, and with a poor quality of life. Consequently, rather then simply being a procedure that some women might consider, vaginal rejuvenation takes on the status of an imperative: indeed, Matlock compares vaginal rejuvenation, and the enhancements the procedure facilitates, to air (as that which makes us thrive). This raises the question of whether and to what extent women are free to choose *not* to undergo surgeries that are being marketed as the obvious (and perhaps only) means by which women (particularly aging women) can maintain culturally prescribed and approved forms of female genitalia and, in turn, femininity.

Further Reading: Apesos, James, Roy Jackson, John R. Miklos, and Robert D. Moore. *Vaginal Rejuvenation: Vaginal/Vulvar Procedures, Restored Femininity.* New York: LM Publishers, 2006; Braun, Virginia, and Celia Kitzinger. "The Perfectible Vagina: Size Matters." *Culture, Health & Sexuality* 3:3 (2001): 263–277; Drenth, Jelto. *The Origin of the World: Science and the Fiction of the Vagina.* London: Reaktion Books, 2005; Green, Fiona J. "From Clitoridectomies to 'Designer Vaginas': The Medical Construction of Heteronormative Female Bodies and Sexuality through Female Genital Cutting." *Sexualities, Evolution and Gender* 7:2 (2005): 153–187; Matlock, David. *Sex by Design.* Los Angeles: Demiurgus Publications, 2004.

Nikki Sullivan

Waist

History of the Corset

The corset is a garment that creates the appearance of a small waist and was historically worn from the sixteenth through the early twentieth centuries by Western women. Corseting or tight lacing involved tightening the garment over time, sometimes so much that it could reshape the skeleton. Spanning nearly five centuries, the corset's reign in Western fashion has been understood as symbolic of culture's deep influence on the female body. Evolving concepts of power and sexuality have always taken expression through female fashion, and the history of the corset reveals both gender and class dynamics.

Often compared to the binding of women's feet in Chinese culture, corsetry for many critics represents a cruel practice undertaken to meet beauty ideals and to control women's bodies. In addition, the corset has been understood as a significant marker of the class of the wearer, suggesting in part the frailty and delicacy of upper-class women and their unsuitability for work. Sociologist Thorstein Veblen (1857–1929) wrote in the early twentieth century that, "it is the use of the corset that provides an absolutely guaranteed signal of female uselessness." For a time, the corset served to symbolize and reinforce the prestige of the ruling class, and women from the sixteenth century to the late nineteenth century rarely challenged the corset. However, some died due to the constriction of corsets, and some pregnant women experienced miscarriage. More common than death, though, were stifled development, permanent deformity, and constant discomfort, all of which hindered even the slightest useful exertion. Eventually, the corset was rejected as an everyday garment, and is now associated with exoticism and fetish dress.

First Appearance and Early Versions

The earliest predecessor to the corset might be traced back to Minoan culture on the island of Crete (ca. 2000–ca. 1400 BCE). Archeological evidence suggests that Minoan women may have painted their faces or wore makeup of some kind, and wore cloths wrapped around the waist under their breasts, which remained exposed. But the corset as we know it has indefinite origins. The earliest references come out of mid-fourteenth-century Europe, with the emergence of a style of dress that exaggerated, even completely falsified, the shape and size of the breasts and waist in an arguably sexualized way. It was not until later in fifteenth-century Spain, however, when a garment referred to as a "body," an armor-like structure consisting of two pieces hinged together at the sides, contoured the female body into a shape that appeared rigid and decidedly unsexual, compressing and even hiding the breasts. It is this version of the corset that we associate with patriarchal tyranny and female disempowerment.

The corset would take many shapes, define many fashions, and display breasts in many various and even contradictory ways over the next few centuries. While it was initially worn exclusively by upper-class women whose mobility was seen as superfluous, it was eventually a garment of the masses, worn by middle-class and even lower-class women to show their upward social mobility and distance from physical labor. The corset was also worn by lower-class women who were trying to attract the attention of newly wealthy industrialists for marriage.

Woman wearing corset, ca. 1899. Courtesy of Library of Congress, LC-USZ62-101143.

During the sixteenth century, the garment, sometimes worn under the clothes, and sometimes as an outergarment, emerged in English and French high society; the rigidity of these versions was achieved by the use of whale bones, wood, ivory, horn, or metal busks. Throughout the next centuries, depending on the fashion and the prevailing mores, corsets were constructed to either enhance or diminish the appearance of breasts.

The evolving and ever-changing corset was, however, almost always consistent with regard to waist minimization. However, one exception existed in fifteenth-century Burgundy. Here, in addition to a wide belt worn beneath the bust to prop it up, women wore a stuffed sack under their clothing to achieve the appearance of a womb-like stomach. At a time when Europe was sparsely populated, this was, perhaps, symbolic of fertility. This enlarged waistline was sharply contrasted by that which the corset was simultaneously popularizing in other parts of Europe.

Veblen in his book *Theory of the Leisure Class* (1899) was the first to articulate the idea that tight lacing was practiced by upper-class women to advertise their leisure, the tiny waist emphasizing the fact that the torso had not grown thick from labor. However, it was primarily the middle classes who used this practice in an effort to mimic the figure of the elites.

The female waist was undoubtedly a symbol of Western prosperity. In addition, because waist minimization was more often than not a partner to breast enhancement, these two exaggerations served to reinforce one another; by comparison with the corseted waist, large, firm, raised breasts and wide hips served to enhance the impression of youth.

At the time of the Renaissance, ancient ideals of naturalism were revived. But while art became overtly classical, the human body was sculpted with complex artifice. Around 1500, female fashion underwent an important structural change in response to the new Italian Renaissance aesthetic of broader, squarer lines. While medieval dress had presented a continuous vertical line, in the early 1500s, the bodice became separated from the skirt. As the skirt became fuller, the bodice grew tighter and more rigid. A new look of was established, which corresponded, according to historian Kunzle, "to the new capitalist ethic, simultaneously expressing power through bulk, and self-restraint through tightness" (Kunzle, 70). Toward the end of the Renaissance, trends in fashion reached new heights of rigidity and severity. The corset was so tight and constricting that it caused woman to faint, and sometimes resulted in deformity and even death.

The eighteenth century and the Enlightenment galvanized the medical profession into a crusade against the corset and the damage caused by tight lacing. The first cries were heard from English philosopher John Locke (1632–1704) in his 1693 treatise *Some Thoughts Concerning Education*. His concerns had mostly to do with children's corsets, but his work was translated into almost every major European language and was a harbinger for a much more general argument against the corset. Over the course of the eighteenth century, a number of texts were written, although for only a limited and predominantly medical audience and published mostly in Latin and German. In 1741, however, Jean-Baptiste Winslow read a paper written in French (the first presentation on this subject given in the vernacular) before the Parisian Academy of the Sciences; Winslow gave a heated diatribe against the corset, discussing in detail the damage to the liver, intestines, kidneys, lungs, and heart caused by forcing the ribs inward. Criticism began to reach a wider audience in Germany in 1743 and in France in 1753 through lengthy encyclopedia entries published in *Zedler's* and *Encyclopedie*, respectively. The first popular criticism was a furious and detailed article written in 1754 by German physician Gottlieb Oelssner, entitled in part: "Philosophical-Moral-Medical Considerations on the several harmful forcible means devised in the interests of pride and beauty by young and adult people of both sexes." Despite the warnings, however, the corset and the body shape it created persisted.

The nineteenth century saw both the democratization of the corset and its eventual downfall. The corset had been prohibitively expensive and impractical for the working classes. Corsets were expensive (the equivalent of perhaps $140 in today's dollars). Moreover, getting dressed was a complicated production. Corsets laced up the back, therefore requiring a maid's assistance, giving this style of dress an aristocratic prestige. Over the course of the nineteenth century, however, a number of advancements made the corset more available to the masses and easier to lace up. In 1823, at the

Exposition Universelle, Jean-Julien Josselin exhibited the first mechanical corset equipped with small pulleys. In 1828, the metal eyelet was invented which both strengthened the corset and greatly simplified the task of hooking it up the back. In 1832, a Swiss tailor named Jean Werly established the first factory to weave seamless corsets, making them distinctly less expensive. In 1840, a system known as "lazy lacing" was developed that truly diminished the need for assistance in dressing.

At the end of the nineteenth century, women's corsets were so tight that their movements were restricted to standing upright and were accompanied by garters, suspenders, false bottoms, and, eventually, artificial breasts. This extreme can be understood as the denouement of the corset's long reign, for a trend was mounting around the turn of the twentieth century to abolish the use of corsets altogether. A number of Eastern European countries actually banned the corset, and the popularity of the garment declined in the West with the rise of the women's liberation movements in the second half of the nineteenth century. Popular beginning in about the 1880s, breast-support garments, precursors to the modern bra, were known in the Victorian era as "emancipation garments" in contrast to the restrictive corset.

Around the turn of the century, Isadora Duncan (1877–1927), an American dancer and communist affiliate, subverted cultural body ideals and fervently decried restrictive forms of dress. She brought the Greek style, characterized by a loose tunic with a high waistline, back into fashion. However, it was not just designers and new trends in popular dictates that caused the decline of the corset. The emergence of middle-class, working woman rendered the restrictive corset impractical. World War I brought this trend to a head. While men were off fighting, women stepped up to fill the posts that they had left behind. These new duties greatly impacted women's dress; hemlines rose and corsets shrank and loosened. Female employment in factories reduced the availability of household servants, and bourgeois women, deprived of dressing assistance, abandoned their complicated style of dress.

Criticisms of the Corset

The corset's popularity was accompanied, as far back as the fifteenth century, by harsh criticism. The corset aroused censure from moralists and religious figures who were offended and scandalized by the sexualization of the breasts, from doctors who were concerned about their extreme physical repercussions, and from early feminists who were concerned by both their symbolic and literal restraint.

Much of the medical literature written against the corset is characterized by a concern for the impediments they posed to natural development, specifically in the case of young girls who were forced into corsets, ironically, to facilitate the development of their posture; however, these fervent opinions against the corset were fraught with disdain for the compliant woman. While corset wearing was certainly, in popular thinking, seen as a female duty, it could alternatively be pitted against women's vanity and disregard for the safety of their wombs. The corseted figure had become such a sign of femininity that women were reluctant to give in to their expanding stomachs during pregnancy. As a result, the corset was responsible for the termination of many pregnancies. From this perspective, the corset became a symbol of anti-maternalism.

In addition, medical critique often conflated indecency with health issues. Physician Oelssner went so far as to claim that the corset was responsible for "toothache, amnesia, and the mumps, as well as colds."

Feminist Response: The Female Body as Cultural Canvass

Perhaps the most vocal members of the anti-corset movement were Jean-Jacques Rousseau (1712–1778), Napoleon Bonaparte (1769–1821), and Pierre-Auguste Renoir (1841–1919). Their fervor was unapologetically misogynistic, driven not by concern for the body of the female but rather for her dedication to her domestic and maternal roles.

However, the nineteenth century began to see a growing feminist voice against the corset. Many women viewed the corset as an obstacle to equal rights, including Susan B. Anthony (1820–1906), who proclaimed, "I can see no business avocation in which woman in her present dress can possibly earn equal wages with man." However, in the 1850s, Elizabeth Cady Stanton (1815–1902) defended fashion as an individual choice and called for a redirection of attention from dress reform to the larger social, economic, and political issues relevant to women's liberation. Paradoxically, tight lacing hit its peak in the 1870s and 1880s, a time when the feminist movement was notably active and productive and when higher education became available to women. The provocative, figure-enhancing style popular during this time was representative of a growing liberalism in attitudes toward sex. The physical dangers of the corset were eclipsed, it appears, by this declaration of female sexuality. In fact, for young, late-Victorian era woman, putting on a corset was somewhat rebellious, embracing its sexual appeal while symbolically rejecting the conservative roles of women of the time.

Nonetheless, the corset did not survive changing women's roles, and was largely replaced by the brassiere, initially called an "emancipation garment" because it was seen to be less restricting than its predecessor.

The Corset as Fetish

In the aftermath of World War I, corsets and other superfluous or impractical garments began to disappear, and by the 1920s, the loose and waistless flapper style was the height of fashion. From this point forward, the corset would resurface in spurts, always drawing on nostalgia for the original item upon which pleasure and pain merged. Currently, the corset is embraced primarily as a fetish object, used in sexualized dress-up and connoting sadomasochistic play. *See also* Breasts: History of the Brassiere.

Further Reading: Barnard, Malcolm. *Fashion as Communication.* New York: Routledge, 1996; Breward, Christopher. *The Culture of Fashion.* New York: St. Martin's Press, 1995; Fontanel, Beatrice. *Support and Seduction: A History of Corsets and Bras.* New

York: Harry N. Abrams, 1997; Kemper, Rachel H. *Costume.* New York: Newsweek Books, 1977; Kunzle, David. *Fashion and Fetishism: A Social History of the Corset, Tight-Lacing and Other Forms of Body-Sculpture in the West.* Lanham, MD: Rowman & Littlefield, 1982; Roach, Mary Ellen, and Joanne Bubotz Eicher. *Dress, Adornment, and the Social Order.* New York: John Wiley & Sons, 1965; Steele, Valerie. *Fetish: Fashion, Sex and Power.* New York: Oxford University Press, 1996; Wilson, Christina. "The History of Corsets," http://ncnc.essortment.com/historyofcors_rmue.htm; Yalom, Marilyn. *A History of the Breast.* New York: Alfred A. Knopf, 1997.

Alana Welch

SELECTED BIBLIOGRAPHY

Abbink, Jon. "Tourism and its Discontents: Suri-Tourist Encounters in Southern Ethiopia." *Social Anthropology* 8, 1 (2000): 1–17.

Abusharaf, Rogaia Mustafa. "Virtuous Cuts: Female Genital Circumcision in an African Ontology." *Differences: A Journal of Feminist Cultural Studies* 12 (2001): 112–140.

Adelman, Marilyn Marx, and Eileen Cahill. *Atlas of Sperm Morphology*. Chicago: ASCP Press Image, 1989.

Ahmed, Leila. *Women and Gender in Islam: Historical Roots of a Modern Debate*. New Haven: Yale University Press, 1992.

Allen, Margaret. *Selling Dreams: Inside the Beauty Business*. New York: Simon & Schuster, 1981.

Almroth, Lars, Vanja Almroth-Berggren, Osman Mahmoud Hassanein, Said Salak Eldin Al-Said, Sharis Siddiq Alamin Hasan, Ulla-Britt Lithell, and Staffan Bergstrom. "Male Complications of Female Genital Mutilation." *Social Science and Medicine* 53 (2001): 1,455–1,460.

Altalbe, Madeline. "Ethnicity and Body Image: Quantitative and Qualitative Analysis." *International Journal of Eating Disorders* 23 (1998): 153–159.

Anderson, Mary M. *Hidden Power: The Palace Eunuchs of Imperial China*. Buffalo, NY: Prometheus, 1990.

Andressen, Michael B. *Spectacles: From Utility to Cult Object*. Stuttgart: Arnoldsche, 1998.

Anzieu, Didier. *The Skin Ego*. New Haven: Yale University Press, 1989.

Aoki, Kenichi. "Sexual Selection as a Cause of Human Skin Color Variation: Darwin's Hypothesis Revisited." *Annals of Human Biology* 29 (2002): 589–608.

Apesos, Jame, Roy Jackson, John R. Miklos, and Robert D. Moore. *Vaginal Rejuvenation: Vaginal/Vulvar Procedures, Restored Femininity*. New York: LM Publishers, 2006.

Arogundade, Ben. *Black Beauty*. London: Pavilion Books, 2000.

Aronowitz, Robert A. *Unnatural History: Breast Cancer and American Society*. Cambridge: Cambridge University Press, 2007.

Atrens, Dale. *The Power of Pleasure*. Sydney: Duffy and Snellgrove, 2000.

Attard, Robert. *Collecting Military Headgear: A Guide to 5000 Years of Helmet History*. Otglen, PA: Schiffer Publishing, 2004.

Bailey, J. Michael, and Aaron S. Greenberg. "The Science and Ethics of Castration: Lessons from the Morse Case." *Northwestern University Law Review* (Summer 1998): 1,225–1,245.

Baker, Robin. *Sperm Wars: The Science of Sex*. Diane Books Publishing Company, 1996.

Baker, Robin, and Mark Bellis. *Human Sperm Competition: Copulation, Masturbation, and Infidelity.* New York: Springer, 1999.
Balsamo, Anne. *Technologies of the Gendered Body: Reading Cyborg Women.* Durham, NC: Duke University Press, 1996.
Barber, Nigel. "The Evolutionary Psychology of Physical Attractiveness: Sexual Selection and Human Morphology," *Ethnology and Sociobiology* 16 (1995): 395–424.
Barker-Benfield, Ben. "The Spermatic Economy: A Nineteenth-Century View of Sexuality." *Feminist Studies* 1:1 (1972): 45–74.
Barnard, Malcolm. *Fashion as Communication.* New York: Routledge, 2002.
Barnett, R. "Historical Keywords: Obesity." *The Lancet* 365, 1843 (2005).
Basow, Susan, and J. Willis. "Perceptions of Body Hair on White Women: Effects of Labeling." *Psychological Reports* 89 (2002): 571–576.
Bell, Kirsten. "Genital Cutting and Western Discourses on Sexuality." *Medical Anthropology Quarterly (New Series)* 19/2 (2005): 125–148.
Berman, Judith C. "Bad Hair Days in the Paleolithic: Modern (Re)Constructions of the Cave Man." *American Anthropologist* 101, 2 (1999): 288–304.
Best, Charles, and Taylor, Norman. *The Living Body: A Text in Human Physiology.* New York: Holt, Rinehart and Winston, 1961.
Bettelheim, Bruno. *Symbolic Wounds: Puberty Rites and the Envious Male.* New York: The Free Press, 1954.
Bianchi, Robert Steven. *Daily Life of the Nubians.* Westport, CT: Greenwood Press, 2004.
Birkhead, Tim, and A. P. Møller, eds. *Sperm Competition and Sexual Selection.* San Diego, CA: Academic Press, 1992.
Björntorp, Per, ed. *International Textbook of Obesity.* Chichester: John Wiley & Sons, 2001.
Blackledge, Catherine. *The Story of V: A Natural History of Female Sexuality.* New Brunswick, NJ: Rutgers University Press, 2003.
Blair, Kirstie. *Victorian Poetry and the Culture of the Heart.* Oxford: Oxford University Press, 2006.
Blank, Hanne. *Virgin: The Untouched History.* New York: Bloomsbury, 2007.
Blistene, Bernard. *Orlan: Carnal Art.* Paris: Editions Flammarion, 2004.
Blum, Virginia L. *Flesh Wounds: The Culture of Cosmetic Surgery.* Berkeley: University of California Press, 2003.
Bohannan, Paul. "Beauty and Scarification amongst the Tiv." *Man* 56 (September 1956): 117–121.
Bordo, Susan. *The Male Body: A New Look at Men in Public and in Private.* New York: Farrar, Strauss and Giroux, 2000.
Bordo, Susan. *Unbearable Weight: Feminism, Western Culture, and the Body.* Berkeley: University of California Press, 1993.
Bossan, Marie-Josephe. *The Art of the Shoe.* New York: Parkstone Press, 2004.
Brack, Datha Clapper. "Displaced: The Midwife by the Male Physician." *Women and Health* 1 (1976): 18–24.
Bradburne, James M., ed. *Blood: Art, Power, Politics, and Pathology.* New York: Prestel, 2002.
Brain, Robert. *Decorated Body.* New York: Harper & Row, 1979.
Brand, Peg Z., ed. *Beauty Matters.* Bloomington: Indiana University Press, 2000.
Braun, Lundy. "Engaging the Experts: Popular Science Education and Breast Cancer Activism." *Critical Public Health* 13 (2003): 191–206.
Braun, Virginia. "In Search of (Better) Sexual Pleasure: Female Genital 'Cosmetic' Surgery." *Sexualities* 8:4 (2005): 407–24.
Braun, Virginia, and Celia Kitzinger. "Telling it Straight? Dictionary Definitions of Women's Genitals." *Journal of Sociolinguistics* 5, 2 (2001): 214–233.
Braun, Virginia, and Celia Kitzinger. "The Perfectible Vagina: Size Matters." *Culture, Health & Sexuality* 3:3 (2001): 263–77.
Braziel, Jana Evans, and Kathleen LeBesco. *Bodies Out of Bounds: Fatness and Transgression.* Berkeley: University of California Press, 2001.

Bremmer, Jan. *The Early Greek Concept of the Soul.* Princeton, NJ: Princeton University Press, 1983.
Breward, Christopher. *The Culture of Fashion.* New York: St. Martin's Press, 1995.
Brodie, Janet Farrell. *Contraception and Abortion in Nineteenth-Century America.* Ithaca, NY: Cornell University Press, 1994.
Bronner, Simon J., ed. *Manly Traditions: The Folk Roots of American Masculinities.* Bloomington: Indiana University Press, 2005.
Bryd, Ayana, and Lori L. Tharps. *Hair Story: Untangling the Roots of Black Hair in America.* New York: St. Martin's Press, 2001.
Bryer, Robin. *The History of Hair: Fashion and Fantasy Down the Ages.* London: Philip Wilson Publishers, 2000.
Budge, E. A. Wallis, trans. *The Egyptian Book of the Dead.* New York: Dover, 1967.
Bulik, Cynthia, Patrick Sullivan, and Kenneth Kendler. "An Empirical Study of the Classification of Eating Disorders." *American Journal of Psychiatry* 157 (2000): 886–895.
Bullough, Vern L., and James A. Brundage, eds. *Handbook of Medieval Sexuality.* New York: Garland, 1996.
Cant, John G. H. "Hypothesis for the Evolution of Human Breasts and Buttocks." *The American Naturalist* 11 (1981): 199–204.
Caplan, Jane, ed. *Written on the Body: The Tattoo in European and American History.* London: Reaktion Books, 2000.
Carlson, Elof Axel. *The Unfit: A History of a Bad Idea.* Woodbury, NY: Cold Spring Harbor Laboratory Press, 2001.
Catlin, George. *O-Kee-Pa: A Religious Ceremony and Other Customs of the Mandan.* Lincoln: University of Nebraska Press, 1976.
Chapman, David L. *Sandow the Magnificent: Eugen Sandow and the Beginnings of Bodybuilding.* Champaign: University of Illinois Press, 1994.
Chelala, C. "An Alternative Way to Stop Female Genital Mutilation." *The Lancet* 352 (1998): 122–126.
Chernin, Kim. *The Obsession: Reflections on the Tyranny of Slenderness.* New York: Harper Perennial, 1994.
Chesser, Barbara Jo. "Analysis of Wedding Rituals: An Attempt to Make Weddings More Meaningful." *Family Relations* 29 (1980): 204–209.
Classen, Albrecht. *The Medieval Chastity Belt: A Myth-Making Process.* New York: Palgrave, 2007.
Cohen, Joseph. *The Penis Book.* Cologne, Germany: Konemann Verlagsgesellschaft, 1999.
Cohen, Meg, and Karen Kozlowski. *Read My Lips: A Cultural History of Lipstick.* San Francisco: Chronicle Books, 1998.
Cohen, Tony. *The Tattoo.* Mosman, NSW: Outback Print, 1994.
Cook, Daniel Thomas, and Susan B. Kaiser. "Betwixt and be Tween: Age Ambiguity and the Sexualization of the Female Consuming Subject," *Journal of Consumer Culture* 2 (2004): 203–27.
Cooper, Charlotte. *Fat and Proud: The Politics of Size.* London: The Women's Press, 1998.
Cornell, Drucilla. *The Imaginary Domain: Abortion, Pornography & Sexual Harassment.* New York: Routledge, 1995.
Corson, Richard. *Fashions in Hair: The First Five Thousand Years.* London: Peter Owen, Tenth Impression, 2005.
Corson, Richard. *Fashions in Makeup: From Ancient to Modern Times.* New York: Peter Owen, 2004.
Craig, Maxine Leeds. *Ain't I a Beauty Queen? Black Women, Beauty and the Politics of Race.* Oxford: Oxford University Press, 2002.
Crapanzano, Vincent. *The Hamadsha: A Study in Moroccan Ethnopsychiatry.* Berkeley: University of California Press, 1973.
Crawford M., and R. Under, eds. *In Our Own Words: Readings on the Psychology of Women and Gender.* New York: McGraw-Hill, 1997.

Crivellato, Enrico, and Domenico Ribatti. "Soul, Mind, Brain: Greek Philosophy and the Birth of Neuroscience." *Brain Research Bulletin* (2007): 327–336.
Cunnington, C. Willett. *The History of Underclothes*. London: M. Joseph, 1951.
Davies, A. "Re-survey of the Morphology of the Nose in Relation to Climate." *Journal of the Royal Anthrolopological Institute of Great Britain and Ireland* 62 (1932): 337–359.
Davis, Kathy. *Dubious Equalities & Embodied Differences: Cultural Studies on Cosmetic Surgery*. Lanham, MD: Rowman & Littlefield, 2003.
Davis, Kathy. *Reshaping the Female Body: the Dilemma of Cosmetic Surgery*. New York: Routledge, 1995.
Dawkins, Richard. *The Selfish Gene*. Oxford: Oxford University Press, 1976.
Delaney, Janice, Mary Jane Lupton, and Emily Toth. *The Curse: A Cultural History of Menstruation*. New York: New American Library, 1977.
Delang, Claudio. *Living at the Edge of Thai Society: The Karen in the Highlands of Northern Thailand*. New York: Routledge, 2003.
DeMello, Margo. *Bodies of Inscription: A Cultural History of the Modern Tattoo Community*. Durham, NC: Duke University Press, 2000.
Denniston, George, Frederick Mansfield Hodges, and Marilyn Fayre Milos. *Male and Female Circumcision: Medical, Legal, and Ethical Considerations in Pediatric Practice*. New York: Springer, 1999.
Devi, Mahasweta. *Breast Stories*. Trans. Gayatri Chakravorty Spivak. Calcutta: Seagull Books, 1997.
Dickinson, Robert Latou. *Human Sex Anatomy*. Baltimore, MD: Williams & Wilkins, 1949.
Dixon, Susan. *The Roman Family*. Baltimore, MD: The Johns Hopkins University Press, 1992.
Donchin, Anne, and Laura Purdy, eds. *Embodying Bioethics: Recent Feminist Advances*. Lanham, MD: Rowman & Littlefield, 1999.
Dowdey, Patrick, and Meifan Zhang, eds. *Threads of Light: Chinese Embroidery and the Photography of Robert Glenn Ketchum*. Los Angeles: UCLA Press, 2002.
Dreger, Alice Domurat. *Hermaphrodites and the Medical Invention of Sex*. Cambridge, MA: Harvard University Press, 1998.
Dreger, Alice Domurat, ed. *Intersex in the Age of Ethics*. Hagerstown, MD: University Publishing Group, 1999.
Drenth, Jelto. *The Origin of the World: Science and the Fiction of the Vagina*. London: Reaktion Books, 2005.
Dundes, Alan, ed. *The Evil Eye: A Case Book*. University of Wisconsin Press: Wisconsin, 1992.
Dutton, Kenneth R. *The Perfectible Body: The Western Ideal of Male Physical Development*. New York: Continuum, 1995.
Ebin, Victoria. *The Body Decorated*. London: Thames & Hudson, 1979.
Eichberg, Sarah L. "Bodies of Work: Cosmetic Surgery and the Gendered Whitening of America." PhD diss., University of Pennsylvania, 1999.
El Guindi, Fadwa. *Veil: Modesty, Privacy and Resistance*. Oxford: Berg, 2000.
Elachalal, Uriel, Barbara Ben-Ami, Rebecca Gillis, and Ammon Brzezinski. "Ritualistic Female Genital Mutilation: Current Status and Future Outlook." *Obstetrical and Gynecological Survey* 52 (1997): 643–651.
Enoch, Jay M. "The Enigma of Early Lens Use." *Technology and Culture* 39, no. 2 (April 1998): 273–291.
Ensler, Eve. *The Vagina Monologues*. New York: Random House, 1998.
Fantham, Elaine, Helene Peet Foley, Natalie Boymel Kampen, and Sarah B. Pomeroy. *Women in the Classical World*. Oxford: Oxford University Press, 1994.
Faris, James C. *Nuba Personal Art*. London: Duckworth, 1972.
Farley, John. *Gametes and Spores: Ideas about Sexual Reproduction 1750–1914*. Baltimore, MD: The Johns Hopkins University Press, 1982.
Farrell-Beck, Jane, and Colleen Gau. *Uplift: The Bra in America*. Philadelphia: University of Pennsylvania Press, 2004.

Fastlicht, Samuel. *Tooth Mutilations and Dentistry in Pre-Columbian Mexico*. Chicago: Quintessence Books, 1976.

Fausto-Sterling, Anne. *Sexing the Body: Gender Politics and the Construction of Sex*. New York: Basic Books, 2000.

Favazza, Armando, R. *Bodies Under Siege: Self-mutilation and Body Modification in Culture and Psychiatry*. Baltimore, MD: The John Hopkins University Press, 1996.

Fildes, Valerie. *Wet Nursing: A History from Antiquity to the Present*. Oxford: Basil Blackwell, 1988.

Fitzherbert, Andrew. *The Palmist's Companion: A History and Bibliography of Palmistry*. Lanham, MD: Rowman & Littlefield, 1992.

Fontanel, Béatrice. *Support and Seduction: The History of Corsets and Bras*. New York: Harry N. Abrams, 1997.

Foucault, Michel. *Herculine Barbin: Being the Recently Discovered Memoirs of a Nineteenth-Century French Hermaphrodite*. New York: Pantheon, 1980.

Fraser, Suzanne. *Cosmetic Surgery, Gender and Culture*. London: Palgrave Macmillan, 2003.

Frater, Lara. *Fat Chicks Rule!: How To Survive In A Thin-centric World*. New York: Gamble Guides, 2005.

Friedman, David M. *A Mind of Its Own: A Cultural History of the Penis*. New York: Free Press, 2001.

Freud, Sigmund. *Three Essays on the Theory of Sexuality*. London: Hogarth Press, 1953 [1905].

Freud, Sigmund. *The Standard Edition of the Complete Psychological Works of Sigmund Freud*. James Strachey, ed. London: Hogarth Press, 1925, 1953.

Gaesser, Glenn. *Big Fat Lies: The Truth about Your Weight and Your Health*. Carlsbad, CA: Gurze Books, 2002.

Gard, Michael, and Wright, Jan. *The Obesity Epidemic: Science, Morality and Ideology*. New York: Routledge, 2005.

Gardner, Howard. *The Mind's New Science: A History of the Cognitive Revolution*. New York: Basic Books, 1985.

Garrett, Valery M. *Traditional Chinese Clothing in Hong Kong and South China, 1840–1980*. New York: Oxford University Press, 1987.

Gaskin, Ina May. *Ina May's Guide to Childbirth*. New York: Bantam, 2003.

Gattuso, John, ed. *Talking to God: Portrait of a World at Prayer*. Milford, NJ: Stone Creek Publications, 2006.

Gibson, Kathleen R., and Tim Ingold. *Tools, Language and Cognition in Human Evolution*. Cambridge: Cambridge University Press, 1995.

Gilman, Sander. "By a Nose: On the Construction of 'Foreign Bodies.'" *Social Epistemology* 13 (1999): 49–58.

Gilman, Sander L. *Making the Body Beautiful: A Cultural History of Aesthetic Surgery*. Princeton, NJ: Princeton University Press, 1999.

Gimlin, Debra. *Body Work: Beauty and Self-Image in American Culture*. Berkeley: University of California Press, 2002.

Glucklich, Ariel. *Sacred Pain: Hurting the Body for the Sake of the Soul*. New York: Oxford University Press, 2001.

Godwin, Gail. *Heart*. New York: HarperCollins, 2001.

Goer, Henci. *Obstetric Myths versus Research Realities: A Guide to the Medical Literature*. Westport, CT: Bergin and Garvey, 1995.

Gordon, Linda. *Woman's Body, Woman's Right: Birth Control in America*. New York: Penguin, 1990.

Gould, Stephen Jay. *The Mismeasure of Man*. W. W. Norton, 1996.

Graham, Ian D. *Episiotomy: Challenging Obstetric Interventions*. Oxford: Blackwell Science, 1997.

Grebowicz, Margret, ed. *Gender After Lyotard*. Albany: State University of New York Press, 2007.

Green, Fiona J. "From Clitoridectomies to 'Designer Vaginas:' The Medical Construction of Heteronormative Female Bodies and Sexuality through Female Genital Cutting." *Sexualities, Evolution and Gender* 7:2 (2005): 153–187.
Green, Monica Helen. "From 'Diseases of Women' to 'Secrets of Women:' The Transformation of Gynecological Literature in the Later Middle Ages," *Journal of Medieval and Early Modern Studies* 30, no. 1 (Winter 2000): 5–39.
Griffen, Joyce. "A Cross-Cultural Investigation of Behavioral Changes at Menopause." *Social Science Journal* 14 (1977): 49–55.
Groning, Karl. *Decorated Skin: A World Survey.* Munich: Frederling and Thaler, 1997.
Guins, Raiford, and Omayra Zaragoza Cruz, eds. *Popular Culture: A Reader.* London: Sage Publications, 2005.
Gullette, Margaret. *Aged by Culture.* Chicago: University of Chicago Press, 2004.
Gulzow, Monte, and Carol Mitchell. "'Vagina Dentata' and 'Incurable Venereal Disease' Legends from the Viet Nam War." *Western Folklore* 39, no. 4 (1980): 306–316.
Gunn, Fenja. *The Artificial Face: A History of Cosmetics.* New York: Hippocrene, 1983.
Guttmacher, Alan. *Pregnancy and Birth: A Book for Expectant Parents.* New York: New American Library, 1962.
Hackett, Earle. *Blood: The Paramount Humour.* London: Jonathan Cape, 1973.
Haddad, Yvonne Yazbeck, Jane I. Smith, and Kathleen M. Moore. *Muslim Women in America: The Challenge of Islamic Identity Today.* New York: Oxford University Press, 2006.
Hagner, Michael. "The Soul and the Brain Between Anatomy and Naturphilosphie in the Early Nineteenth Century." *Medical History* 36 (1992): 126–137.
Haiken, Elizabeth. *Venus Envy: A History of Cosmetic Surgery.* Baltimore, MD: The Johns Hopkins University Press, 1997.
Haraway, Donna J. *Primate Visions: Gender, Race, and Nature in the World of Modern Science.* New York: Routledge, 1989.
Harrington, Anne. *Medicine, Mind, and the Double Brain: A Study in Nineteenth-Century Thought.* Princeton, NJ: Princeton University Press, 1987.
Harrington, Charles. "Sexual Differentiation in Socialization and Some Male Genital Mutilations." *American Anthropologist (New Series)* 70/5 (1968): 951–956.
Harris, W. V. "Child-Exposure in the Roman Empire," *The Journal of Roman Studies* 84 (1994): 1–22.
Harvey, Adia. "Becoming Entrepreneurs: Intersections of Race, Class, and Gender at the Black Beauty Salon." *Gender and Society* 19, 6 (2005): 789–808.
Henschen, Folke. *The Human Skull: A Cultural History.* New York: Frederick A. Praeger, 1966.
Heselt van Dinter, Maarten. *The World of Tattoo: An Illustrated History.* Amsterdam: KIT Publishers, 2005.
Hewitt, Kim. *Mutilating the Body: Identity in Blood and Ink.* Bowling Green, OH: Bowling Green University Popular Press, 1997.
Hirschmann, Nancy J. "Western Feminism, Eastern Veiling, and the Question of Free Agency." *Constellations: An International Journal of Critical and Democratic Theory* 5 (1998): 345–369.
Hite, Shere. *The Hite Report: a Nationwide Study on Female Sexuality.* New York: Macmillan, 1976.
Hobson, Janell. *Venus in the Dark: Blackness and Beauty in Popular Culture.* New York: Routledge, 2005.
Hogbin, Ian. *The Island of Menstruating Men.* London: Chandler, 1970.
Hopkins, E. Washburn. "The Sniff-Kiss in Ancient India." *Journal of the American Oriental Society* 28 (1907): 120–134.
Houck, Judith A. *Hot and Bothered: Women, Medicine, and Menopause in Modern America.* Cambridge, MA: Harvard University Press, 2006.
Houppert, Karen. *The Curse: Confronting the Last Taboo: Menstruation.* London: Profile Books, 2000.

Howard, David. *The Last Filipino Head Hunters*. Last Gasp, 2001.
Hunter, Margaret. "'If You're Light, You're Alright': Light Skin Color as Social Capital for Women of Color." *Gender and Society* 61 (2002): 175–193.
Hurston, Zora Neale. *Their Eyes Were Watching God*. Urbana: University of Illinois Press, 1978 [1937].
Iiardi, Vincent. "Eyeglasses and Concave Lenses in Fifteenth-Century Florence and Milan: New Documents." *Renaissance Quarterly* 29, no. 3 (1976): 341–360.
Ince, Kate. *Orlan: Millennial Female*. Oxford: Berg Publishers, 2000.
Inckle, Kay. *Writing on the Body? Thinking Through Gendered Embodiment and Marked Flesh*. Newcastle-Upon-Tyne: Cambridge Scholars Publishing, 2007.
Inhorn, Marcia, and Frank van Balen, eds. *Infertility Around the Globe: New Thinking on Childlessness, Gender and Reproductive Technologies*. Berkeley: University of California Press, 2002.
Jabet, George. *Notes on Noses*. London: Harrison and Sons Publishing, 1852.
Jablonski, Nina. "The Evolution of Human Skin and Skin Color." *Annual Review of Anthropology* 33 (2004): 585–623.
Jackson, Beverley. *Splendid Slippers: A Thousand Years of an Erotic Tradition*. Berkeley, California: Ten Speed Press, 2000.
Jackson, Bruce. "Vagina Dentata and Cystic Teratoma." *The Journal of American Folklore* 84, no. 333 (1971): 341–342.
Jager, Eric. *The Book of the Heart*. Chicago: University of Chicago Press, 2001.
Jeffreys, Sheila. *Beauty and Misogyny: Harmful Cultural Practices in the West*. New York: Routledge, 2005.
Johnson, Sonali. "The Pot Calling the Kettle Black? Gender-Specific Health Dimensions of Colour Prejudice in India." *Journal of Health Management,* vol. 4, no. 2 (2002): 215–227.
Johnston, David. *Roman Law in Context*. Cambridge: Cambridge University Press, 1999.
Jones, Meredith. *Skintight: A Cultural Anatomy of Cosmetic Surgery*. Oxford: Berg Publishers, 2008.
Kaplan, David E., and Alec Dubro. *Yakuza: Japan's Criminal Underworld*. Berkeley: University of California Press, 2003.
Kaplan, Jonas, Joshua Freedman, and Marco Iacoboni. "Us Versus Them: Political Attitudes and Party Affiliation Influence Neural Response to Faces of Presidential Candidates." *Neuropsychologia* 45.1 (2007): 55–64.
Kasper, Anne S., and Susan J. Ferguson, eds. *Breast Cancer: Society Shapes an Epidemic*. New York: St. Martin's Press, 2000.
Kaw, Eugenia. "Medicalization of Racial Features: Asian American Women and Cosmetic Surgery" *Medical Anthropology Quarterly* 7 (1993): 74–89.
Keane, Helen. *What's Wrong with Addiction?* Carlton South: Melbourne University Press, 2002.
Keele, Kenneth. *Leonardo da Vinci's Elements of the Science of Man*. New York: Academic Press, 1983.
Kelley, Mark A., and Clark Spencer Larsen, eds. *Advances in Dental Anthropology*. New York: Wiley-Liss, 1991.
Kennedy, Carolee G. "Prestige Ornaments: The Use of Brass in the Zulu Kingdom" *African Arts,* vol. 24, no. 3 (1991): 50–55.
Kessler, Suzanne J. *Lessons from the Intersexed*. Piscataway, NJ: Rutgers University Press, 1998.
Kevles, Bettyann. *Naked to the Bone: Medical Imaging in the Twentieth Century*. Reading, MA: Addison-Wesley, 1998.
Kiefer, Otto. *Sexual Life in Ancient Rome*. London: Routledge, 1934.
Kimmel, Michael S. "Men's Responses to Feminism at the Turn of the Century." *Gender and Society* 1:3 (1987): 261–283.
King, Helen. *Hippocrates' Woman: Reading the Female Body in Ancient Greece*. New York: Routledge, 1998.
Kinsey, Alfred, Wardell Pomeroy, Clyde Martin, and Paul Gebhard. *Sexual Behavior in the Human Female*. Bloomington: Indiana University Press, 1998 [1953].

Kirkup, John. *A History of Limb Amputation*. New York: Springer Publishing, 2006.
Klapisch-Zuber, Christiane, ed. *A History of Women: Silences of the Middle Ages*. Cambridge, MA: The Belknap Press of Harvard University Press, 1992.
Klawiter, Maren. *The Biopolitics of Breast Cancer: Changing Cultures of Disease and Activism*. Minneapolis: University of Minnesota Press, 2008.
Ko, Dorothy. *Cinderella's Sisters: A Revisionist History of Footbinding*. Berkeley: University of California Press, 2005.
Ko, Dorothy. *Every Step a Lotus: Shoes for Bound Feet*. Berkeley: University of California Press, 2001.
Kulick, Don. *Fat: The Anthropology of an Obsession*. New York: Tarcher, 2005.
Kulick, Don. *Travesti: Sex, Gender, and Culture among Brazilian Transgendered Prostitutes*. Chicago: University of Chicago Press, 1998.
Kunzle, David. *Fashion and Fetishism: A Social History of the Corset, Tight-Lacing and Other Forms of Body-Sculpture in the West*. Lanham, MD: Rowman & Littlefield, 1982.
Kushner, Rose. *Breast Cancer: A Personal History and an Investigative Report*. New York: Harcourt Brace Jovanovich, 1975.
Kwass, Michael. "Big Hair: A Wig History of Consumption in Eighteenth-Century France." *American Historical Review* June (2006): 631–659.
Lam, Samuel. "Edward Talbot Ely: Father of Aesthetic Otoplasty." *Archives of Facial Plastic Surgery* 6, 1 (2004): 64.
Laqueur, Thomas W. *Making Sex: Body and Gender from the Greeks to Freud*. Cambridge, MA: Harvard University Press, 1990.
Laqueur, Thomas W. *Solitary Sex: A Cultural History of Masturbation*. Cambridge, MA: Zone Books, 2003.
Larrat, Shannon. *ModCon: The Secret World of Extreme Body Modification*. Toronto: BME Books, 2002.
Laws, Sophie. *Issues of Blood: The Politics of Menstruation*. London: Macmillan Press, 1990.
LeBesco, Kathleen. *Revolting Bodies? The Struggle to Redefine Fat Identity*. Amherst: University of Massachusetts Press, 2004.
Leddick, David. *The Male Nude*. Cologne, Germany: Taschen, 1998.
Leeds Craig, Maxine. *Ain't I a Beauty Queen: Black Women, Beauty, and the Politics of Race*. New York: Oxford University Press, 2002.
Lefkowitz, Mary R., and Maureen B. Fant. *Women's Life in Greece and Rome: A Source Book in Translation*. 2nd ed. Baltimore, MD: The Johns Hopkins University Press, 1992.
Lehman, Peter. *Running Scared: Masculinity and the Representation of the Male Body*. Philadelphia: Temple University Press, 1993.
Leong, Samuel, and Paul White. "A Comparison of Aesthetic Proportions Between the Healthy Caucasian Nose and the Aesthetic Ideal." *Journal of Plastic, Reconstructive and Aesthetic Surgery* 59 (2006): 248–252.
Lester, Paul Martin, and Susan Dente Ross. "Body Hair Removal: The 'Mundane' Production of Normative Femininity." *Sex Roles* 52 (2005): 399–406.
Levene, John R. "Sir George Biddell Airy, F. R. S. (1801–1892) and the Discovery and Correction of Astigmatism," *Notes and Records of the Royal Society of London* 21, no. 2 (December 1966): 180–199.
Levenkron, Steven. *Cutting: Understanding and Overcoming Self-Mutilation*. New York: W. W. Norton & Company, 1998.
Levine, Laurie. *The Drumcafe's Traditional Music of South Africa*. Johannesburg: Jacana Media, 2005.
Levy, Howard S. *The Lotus Lovers: The Complete History of the Curious Erotic Custom of Footbinding in China*. Buffalo, NY: Prometheus Books, 1991.
Lewis, J. M. "Caucasion Body Hair Management: A Key to Gender and Species Identification in U. S. Culture." *Journal of American Culture* 10, (1987): 7–18.
Lewis, Paul. *Peoples of the Golden Triangle*. London: Thames & Hudson, 1998.
Licht, Hans. *Sexual Life in Ancient Greece*. London: Routledge, 1932.

Liddell, Scott K. *Grammar, Gesture, and Meaning in American Sign Language*. Cambridge: Cambridge University Press, 2003.

Lloyd, Jillian, Naomi Crouch, Catherine Minto, Lih-Mei Liao, and Sarah Crieghton. "Female Genital Appearance: 'Normality' Unfolds." *British Journal of Obstetrics and Gynaecology* 112, 5 (2005): 643.

Longrigg, James. *Greek Medicine From the Heroic to the Hellenistic Age: A Sourcebook*. New York: Routledge, 1998.

Lorde, Audre. *The Cancer Journals*. Argyle, NY: Spinsters, Ink, 1980.

Louderback, Lewellyn. *Fat Power: Whatever You Weigh Is Right*. New York: Hawthorn Books, 1970.

Luciano, Lynne. *Looking Good: Male Body Image in Modern America*. New York: Hill and Wang, 2002.

Lutkehaus, Nancy C., and Paul B. Roscoe, eds. *Gender Rituals: Female Initiation in Melanesia*. New York: Routledge, 1995.

Maddox, Keith, and Stephanie Gray. "Cognitive Representations of Black Americans: Reexploring the Role of Skin Tone." *Personality and Social Psychology Bulletin* 28 (2002): 250–259.

Maher, Vanessa. *The Anthropology of Breast-Feeding: Natural Law or Social Construct*. Oxford: Berg Publishers, 1992.

Maines, Rachel P. *The Technology of Orgasm: "Hysteria," the Vibrator, and Women's Sexual Satisfaction*. Baltimore, MD: The Johns Hopkins University Press, 1999.

Manniche, Lise. *Sexual Life in Ancient Egypt*. London: Kegan Paul International, 1987.

Mannix, Daniel P. *The History of Torture*. Sutton Publishing, 2003.

Markus, Ruth. "Surrealism's Praying Mantis and Castrating Woman." *Woman's Art Journal* 21, no. 1 (2000): 33–39.

Marmon, Shaun E. *Eunuchs and Sacred Boundaries in Islamic Society*. New York: Oxford University Press, 1995.

Marsh, Margaret, and Ronner, Wanda. *The Empty Cradle: Infertility in America from Colonial Times to the Present*. Baltimore, MD: The Johns Hopkins University Press, 1996.

Marshall, Harry I. *The Karen of the Burma*. Bangkok: White Lotus, 1997 [1922].

Martin, Emily. *The Woman in the Body: A Cultural Analysis of Reproduction*. Boston, MA: Beacon Press, 1987.

Marwick, Arthur. *Beauty in History: Society, Politics and Personal Appearance c. 1500 to the Present*. London: Thames & Hudson, 1988.

Mascetti, Daniela, and Amanda Triossi. *Earrings: From Antiquity to the Present*. London: Thames & Hudson, 1990.

Mashour, George, Erin Walker, and Robert Martuza. "Psychosurgery: Past, Present, and Future." *Brain Research Reviews* 48:3 (2005): 409–419.

Matlock, David. L. *Sex by Design*. Los Angeles: Demiurgus Publications, 2004.

McAndrew, Sue, and Tony Warne. "Cutting Across Boundaries: A Case Study Using Feminist Praxis to Understand the Meanings of Self-Harm." *International Journal of Mental Health Nursing* 14 (2005): 172–180.

McCloud, Aminah. "American Muslim Women and U.S. Society." *Journal of Law and Religion* 12 (1995–1996): 51–59.

McKinnon, John. *Highlanders of Thailand*. Oxford: Oxford University Press, 1983.

McLaren, Angus. *Impotence: A Cultural History*. Chicago: University of Chicago Press, 2007.

McNeil, Peter, and Giorgio Riello. *Shoes: A History from Sandals to Sneakers*. New York: Berg Publishers, 2006.

Mercer, Kobena. "Black Hair/Style Politics." *New Formations* 3 (1987): 33–54.

Meyer, Melissa, L. *Thicker Than Water: The Origins of Blood as Symbol and Ritual*. New York and London: Routledge, 2005.

Miller, Heather Lee. "Sexologists Examine Lesbians and Prostitutes in the United States, 1840–1940." *NWSA Journal* 12, 3 (2000): 67–91.

Millstead, Rachel, and Hannah Frith. (2003) "Being Large Breasted: Women Negotiating Embodiment." *Women's Studies International Forum* 26:5 (2003): 455–465.

Minois, Georges. *History of Old Age: From Antiquity to the Renaissance*, trans. Sarah Hanbury Tenison. Chicago: Chicago University Press, 1989.

Mirante, Edith. *Down the Rat Hole: Adventures Underground on Burma's Frontiers*. Hong Kong: Orchid Press, 2005.

Mitchell, Marilyn Hall. "Sexist Art Criticism: Georgia O'Keeffe, A Case Study." *Signs* 3, 3 (1978): 681–687.

Moldonado, Tomás. "Taking Eyeglasses Seriously." *Design Issues* 17, no. 4 (Autumn 2001): 32–43.

Money, John, and Anke A. Ehrhardt. *Man & Woman, Boy & Girl: The Differentiation and Dimorphism of Gender Identity from Conception to Maturity*. Baltimore, MD: The Johns Hopkins University Press, 1972.

Montagu, Ashley. *Coming into Being amongst the Australian Aboriginies*. 2nd ed. London: Routledge, 1974.

Moore, Lisa Jean, and Adele E. Clarke. "Clitoral Conventions and Transgressions: Graphic Representations of Female Genital Anatomy, c. 1900–1991." *Feminist Studies* 21:2 (1995): 255–301.

Moore, Lisa Jean, and Adele E. Clarke. "The Traffic in Cyberanatomies: Sex/Gender/Sexualities in Local and Global Formations." *Body and Society* 7:1 (2001): 57–96.

Moore, Pamela L., ed. *Building Bodies*. New Brunswick, NJ: Rutgers University Press, 1997.

Morgan, Kathryn Pauly. "Women and the Knife: Cosmetic Surgery and the Colonization of Women's Bodies" *Hypatia*, vol. 6, no. 3 (1991): 25–53.

Morra, Joanne, and Smith, Marquard. *The Prosthetic Impulse: from a Posthuman Present to a Biocultural Future*. Cambridge: MIT Press, 2006.

Morris, Desmond. *The Naked Woman: A Study of the Female Body*. New York: Thomas Dunne Books, 2004.

Moss, Madonna. "George Catlin Among the Nayas: Understanding the Practice of Labret Wearing on the Northwest Coast." *Ethnohistory*, vol. 46, no 1. (1999): 31–65.

Murphy, Robert. "Social Distance and the Veil." *American Anthropologist* 66 (1964): 1,257–1,274.

Musafar, Fakir. *Spirit + Flesh*. Santa Fe, NM: Arena Editions, 2002.

Nash, Joycelyn D. *What Your Doctor Can't Tell You About Cosmetic Surgery*. Oakland, CA: New Harbinger Publications, 1995.

Nuttall, Sarah. "Girl Bodies." *Social Text* 22.1 (2004): 17–33.

Nutton, Vivian. *Ancient Medicine*. New York: Routledge, 2004.

Oakes, Jill, and Rick Riewe. *Our Boots: An Inuit Women's Art*. London: Thames & Hudson, 1995.

Oakley, John H., and Rebecca H. Sinos. *The Wedding in Ancient Athens*. Madison: The University of Wisconsin Press, 1993.

O'Bryan, C. Jill. *Carnal Art: Orlan's Refacing*. Minneapolis: University of Minnesota Press, 2005.

Orbach, Susie. *Fat Is a Feminist Issue*. London: Arrow Books, 1978.

Paffrath, James D., ed. *Obsolete Body/Suspensions/Stelarc*. Davis, CA: J. P. Publications, 1984.

Paley, Maggie. *The Book of the Penis*. New York: Grove Press, 1999.

Palmer, Gabrielle. *The Politics of Breastfeeding*. 2nd ed. London: Pandora Press, 1993.

Parens, Erik, ed. *Surgically Shaping Children: Technology, Ethics, and the Pursuit of Normality*. Baltimore, MD: The John Hopkins University, 2006.

Paterson, Dennis. "Leg-Lengthening Procedures: A Historical Review." *Clinical Orthopaedics and Related Research* 250 (1990): 27–33.

Peiss, Kathy. *Hope in a Jar: The Making of America's Beauty Culture*. New York: Henry Holt and Company, 1998.

Persaud, Walter. "Gender, Race and Global Modernity: A Perspective for Thailand." *Globalizations*, vol. 24, no. 2 (September 2005): 210–227.
Pfeffer, Naomi. *The Stork and the Syringe: A Political History of Reproductive Medicine.* Cambridge: Polity Press, 1993.
Phillips, K. A., R. L. O'Sullivan, and H. G. Pope. "Muscle Dysmorphia." *Journal of Clinical Psychiatry* 58 (1997): 361.
Ping, Wang. *Aching for Beauty: Footbinding in China.* Minneapolis: University of Minnesota Press, 2000.
Pinto-Correia, Clara. *The Ovary of Eve: Egg and Sperm and Preformation.* Chicago: University of Chicago Press, 1997.
Pitts, Victoria. *In the Flesh: the Cultural Politics of Body Modification.* New York: Palgrave, 2003.
Pitts-Taylor, Victoria. *Surgery Junkies: Wellness and Pathology in Cosmetic Culture.* New Brunswick, NJ: Rutgers University Press, 2007.
Polhemus, Ted. *Hot Bodies, Cool Styles: New Techniques in Self-Adornment.* London: Thames & Hudson, 2004.
Pope, Harrison G., Katharine A. Phillips, and Roberto Olivardia. *The Adonis Complex: The Secret Crisis of Male Body Obsession.* New York: Free Press, 2000.
Porter, Roy. *Flesh in the Age of Reason.* New York: W. W. Norton & Co, 2004.
Porter, Roy. *The Greatest Benefit to Mankind: A Medical History of Humanity.* New York: W. W. Norton, 1998.
Post, Stephen G., and Binstock, Robert H. *The Fountain of Youth: Cultural, Scientific, and Ethical Perspectives on a Biomedical Goal.* Oxford University Press, Oxford, 2004.
Potts, Laura K., ed. *Ideologies of Breast Cancer: Feminist Perspectives.* London: Macmillan, 2000.
Powell, J. H. "'Hook-Swinging' in India. A Description of the Ceremony, and an Enquiry into Its Origin and Significance." *Folklore* 25/2 (1914): 147–197.
Preves, Sharon. *Intersex and Identity: The Contested Self.* New Brunswick, NJ: Rutgers University Press, 2003.
Radley, Alan, and Susan E. Bell. "Artworks, Collective Experience and Claims for Social Justice: The Case of Women Living with Breast Cancer." *Sociology of Health & Illness* 29 (2007): 366–390.
Raitt, Jill. "The 'Vagina Dentata' and the 'Immaculatus Uterus Divini Fontis.'" *Journal of the American Academy of Religion* 48, no. 3 (1980): 415–431.
Read, J., and J. Bartkowski. "To Veil or Not to Veil? A Case Study of Identity Negotiation among Muslim Women in Austin, Texas." *Gender and Society* 14 (2000): 395–417.
Reis, Elizabeth. "Impossible Hermaphrodites: Intersex in America, 1620–1960." *The Journal of American History* September (2005): 411–441.
Rhodes, Gillian, and Leslie, A. Zebrowitz, eds. *Facial Attractiveness: Evolutionary, Cognitive, and Social Perspectives.* Westport, CT: Ablex, 2002.
Riddle, John M. *Contraception and Abortion from the Ancient World to the Renaissance.* Cambridge, MA: Harvard University Press, 1994.
Riddle, John M. *Eve's Herbs: A History of Contraception and Abortion in the West.* Cambridge, MA: Harvard University Press, 1997.
Riley, Richard Lee. *Living with a Below-Knee Amputation: A Unique Insight from a Prosthetist/Amputee.* Thorofare: Slack, 2006.
Roberts, Dorothy. *Killing the Black Body: Race, Reproduction, and the Meaning of Liberty.* New York: Pantheon Books, 1997.
Roome, Loretta. *Mehndi: The Timeless Art of Henna Painting.* New York: St. Martin's Griffin, 1998.
Ropp, Paul S., ed. *Heritage of China: Contemporary Perspectives on Chinese Civilization.* Berkeley: University of California Press, 1990.
Rosenbaum, Michael. *The Fighting Arts: Their Evolution from Secret Societies to Modern Times.* Jamaica Plain, MA: YMAA Publication Center, 2002.
Rosenthal, William J. *Spectacles and Other Vision Aids: A History and Guide to Collecting.* San Francisco: Norman Publishing, 1996.

Roth, Melissa. *The Left Stuff: How the Left-handed Have Survived and Thrived in a Right-handed World.* New York: M. Evans and Company, 2005.
Rothman, Barbara Katz. *Recreating Motherhood.* New Brunswick, NJ: Rutgers University Press, 2000.
Rousseau, Jérôme. *Kayan Religion.* Leiden, Netherlands: KITLV Press, 1998.
Rubin, Arnold, ed. *Marks of Civilization.* Los Angeles: Museum of Cultural History, UCLA, 1988.
Rush, John. *Spiritual Tattoo: A Cultural History of Tattooing, Piercing, Scarification, Branding, and Implants.* Berkeley, CA: Frog Ltd., 2005.
Russell, John. "Race and Reflexivity: The Black Other in Contemporary Japanese Mass Culture." *Cultural Anthropology*, vol. 6, no. 1 (1991): 3–25.
Sarwer, David B., Jodi E. Nordmann, and James D. Herbert. "Cosmetic Breast Augmentation: A Critical Overview." *Journal of Women's Health & Gender-Based Medicine*, vol. 9, issue 8 (2000): 843–56.
Saetnan, Anne, Nelly Oudshoorn, and Marta Kirejczyk. *Bodies of Technology: Women's Involvement with Reproductive Medicine.* Columbus: Ohio State University Press, 2000.
Sault, Nicole, ed. *Many Mirrors: Body Image and Social Relations.* New Brunswick, NJ: Rutgers University Press, 1994.
Saville, Marshall H. "Pre-Columbian Decoration of the Teeth in Ecuador. With Some Account of the Occurrence of the Custom in Other Parts of North and South America." *American Anthropologist (New Series)* 15/3 (1913): 377–394.
Schiebinger, Londa, ed. *Feminism and the Body.* New York: Oxford University Press, 2000.
Schiebinger, Londa. "The Anatomy of Difference: Race and Sex in Eighteenth-Century Science." *Eighteenth-Century Studies*, vol. 23, no. 4 (1990): 387–405.
Schwartz, Hillel. *Never Satisfied: A Cultural History of Diets, Fantasies and Fat.* New York: The Free Press, 1986.
Scott, Margaret. *A Visual History of Costume: The Fourteenth & Fifteenth Centuries.* London: B. T. Batsford Ltd, 1986.
Sechzer, Jeri Altneu. "'Islam and Woman: Where Tradition Meets Modernity': History and Interpretations of Islamic Women's Status." *Sex Roles* 51 (2004): 263–272.
Seeger, Anthony. "The Meaning of Body Ornaments: A Suya Example." *Ethnology*, vol. 14, no. 3 (1975): 211–224.
Sevely, Josephine Lowndes. *Eve's Secrets: A New Theory of Female Sexuality.* New York: Random House, 1987.
Shail, Andrew, and Gillian Howie, eds. *Menstruation: A Cultural History.* New York: Palgrave, 2005.
Shankar, Wendy. *The Fat Girl's Guide to Life.* New York: Bloomsbury, 2004.
Sherrow, Victoria. *Encyclopedia of Hair: A Cultural History.* Westport, CT: Greenwood Press, 2006.
Shirazi, Faegheh. *The Veil Unveiled: The Hijab in Modern Culture.* Gainesville: University Press of Florida, 2001.
Shohat, Ella, ed. *Talking Visions: Multicultural Feminism in a Transnational Age.* Cambridge, MA: MIT Press, 1998.
Shuttle, Penelope, and Peter Redgrove. *The Wise Wound: Menstruation and Everywoman.* 2nd ed. London: Marion Boyars, 1999.
Silverstein, Alvin. *Human Anatomy and Physiology.* New York: John Wiley and Sons, 1980.
Sims, Michael. *Adam's Navel: A Natural and Cultural History of the Human Form.* Viking: New York, 2003.
Singer, Philip, and Daniel E. Desole. "The Australian Subincision Ceremony Reconsidered: Vaginal Envy or Kangaroo Bifid Penis Envy." *American Anthropologist (New Series)* 69/3/4 (1967): 355–8.
Smith, Robert L., ed. *Sperm Competition and the Evolution of Animal Mating Systems.* New York: Academic Press, 1984.

Solovay, Sondra. *Tipping the Scales of Justice: Fighting Weight-Based Discrimination*. Amherst, NY: Prometheus Books, 2000.

Spallone, Pat, and Deborah Steinberg. *Made To Order: The Myth of Reproductive and Genetic Progress*. Oxford: Pergamon Press, 1987.

Spark, Richard F. *The Infertile Male*. New York: Springer, 1988.

Spencer, Paul. *The Maasai of Matapato*. New York: Routledge, 2004.

Spindler, Konrad. *Man in the Ice*. New York: Harmony Books, 1994.

Spitzack, Carole. *Confessing Excess: Women and the Politics of Body Reduction*. Albany, NY: State University of New York Press, 1990.

Starr, Douglas. *Blood: An Epic History of Medicine and Commerce*. London: Warner Books, 2000.

Steele, Valerie. *Fetish: Fashion, Sex & Power*. New York: Oxford University Press, 1996.

Steinbach, Ronald D. *The Fashionable Ear*. New York: Vantage Press, 1995.

Stinson, Susan. *Venus of Chalk*. Ann Arbor, MI: Firebrand Books, 2004.

Stolberg, Michael. "A Woman's Hell? Medical Perceptions of Menopause in Preindustrial Europe." *Bulletin of the History of Medicine* 73.3 (1999): 404–428.

Sullivan, Nikki. "'The Price to Pay for Our Common Good:' Genital Modification and the Somatechnologies of Cultural (In)Difference." *Social Semiotics* 17:3 (2007).

Sullivan, Nikki. "'It's as Plain as the Nose on his Face': Michael Jackson, Body Modification, and the Question of Ethics." *Scan: Journal of Media Arts Culture* 3:1 (2004).

Sullivan, Nikki. *Tattooed Bodies*. Westport, CT: Praeger, 2001.

Synnott, Anthony. "Shame and Glory: A Sociology of Hair." *The British Journal of Sociology* 38 (1987): 381–413.

Taylor, Gary. *Castration: An Abbreviated History of Western Manhood*. New York: Routledge, 2000.

Thesander, Marianne. *The Feminine Ideal*. London: Reaktion Books, 1997.

Thomas, Pattie. *Taking Up Space: How Eating Well & Exercising Regularly Changed My Life*. Nashville, TN: Pearlsong Press, 2005.

Throsby, Karen. *When IVF Fails: Feminism, Infertility and the Negotiation of Normality*. Basingstoke: Palgrave, 2004.

Tomlinson, Stephen. *Head Masters: Phrenology, Secular Education, and Nineteenth-Century Social Thought*. Tuscaloosa: University of Alabama Press, 2005.

Tsunenari, S., T. Idaka, M. Kanda, and Y. Koga. "Self-mutilation. Plastic Spherules in Penile Skin in Yakuza, Japan's Racketeers." *American Journal of Forensic Medical Pathology* 3/2 (September 1981): 203–7.

Turner, Terence. "Social Body and Embodied Subject: Bodiliness, Subjectivity, and Sociality Among the Kayapo." *Cultural Anthropology*, vol. 10, no. 2 (1995): 142–170.

Turton, David. "Lip-plates and 'The People Who Take Photographs': Uneasy Encounters Between Mursi and Tourists in Southern Ethiopia." *Anthropology Today*, vol. 20, no. 3 (2004): 3–8.

Vale, V., and Andrea Juno, eds. *Modern Primitives*. San Francisco: Re/Search, 1989.

Valentine, Bill. *Gang Intelligence Manual: Identifying and Understanding Modern-Day Violent Gangs in the United States*. Boulder, CO: Paladin Press, 1995.

Vitols, Astrid. *Dictionnaire des Lunettes, Historique et Symbolique d'un Objet Culturel*. Paris: Éditions Bonneton, 1994.

Vlahos, Olivia. *Body: The Ultimate Symbol*. New York: J. B. Lippincott, 1979.

Voda, Ann M. *Changing Perspectives on Menopause*. Austin: University of Texas Press, 1982.

Wallerstein, Edward. *Circumcision: An American Health Fallacy*. New York: Springer, 1980.

Warnke, Georgia. "Intersexuality and the Categories of Sex." *Hypatia* 16 (2001): 126–137.

Wassersug, Richard J., Sari A. Zelenietz, and G. Farrell Squire. "New Age Eunuchs: Motivation and Rationale for Voluntary Castration." *Archives of Sexual Behavior* 33, no. 5 (2004): 433–442.
Webber, Sara. "Cutting History, Cutting Culture: Female Circumcision in the United States." *The American Journal of Bioethics* 3, 2 (2003): 65–66.
Weideger, Paula. *Female Cycles.* London: The Women's Press, 1978.
Weitz, Rose, ed. *The Politics of Women's Bodies.* Oxford: Oxford University Press, 1998.
Wheeler, Roxanne. "The Complexion of Desire: Racial Ideology and Mid-Eighteenth-Century British Novels." *Eighteenth-Century Studies* 32 (1999): 309–332.
White, Elizabeth. "Purdah." *Frontiers: A Journal of Women Studies* 2 (1977): 31–42.
Wille, Reinhard, and Klaus M. Beier. "Castration in Germany." *Annals of Sex Research* 2 (1989): 105–109.
Winkel, Eric. "A Muslim Perspective on Female Circumcision." *Women and Health* 23 (1995): 1–7.
Winkler, Wolf, ed. *A Spectacle of Spectacles.* Leipzig: Edition Leipzig, 1988.
Winterich, Julie A. "Sex, Menopause, and Culture: Sexual Orientation and the Meaning of Menopause for Women's Sex Lives." *Gender and Society* 17, 4 (2003): 627–642.
Wolf, Naomi. *The Beauty Myth.* New York: William Morrow, 1991.
Yalom, Marilyn. *A History of the Breasts.* New York: Alfred A. Knoff, 1997.
Yalom, Marilyn. *A History of the Wife.* New York: HarperCollins, 2001.
Young, Iris Marion. *On Female Body Experience: 'Throwing Like a Girl' and Other Essays.* Oxford: Oxford University Press, 2005.
Young, Robert M. *Mind, Brain, and Adaptation in the Nineteenth Century: Cerebral Localization and Its Biological Context from Gall to Ferrier.* New York: Oxford University Press, 1990.
Yung, Judy. *Unbound Voices: A Documentary History of Chinese Women in San Francisco.* Berkeley: University of California Press, 1999.
Zimmer, Carl. *Soul Made Flesh: The Discovery of the Brain—and How It Changed the World.* London: Free Press, 2004.

INDEX

NOTE: Page numbers in bold indicate main entries in the encyclopedia; page numbers followed by *f* indicate illustrations.

5-Alpha reductase deficiency (5-ARD), intersex condition, 218

Abdomen, **1–3**
Abdominal etching, 193
Abdominal muscles, abdominoplasty and, 1–3
Abdominoplasty, 1–3, 153, 192
Aboriginal tribes, Australia, subincision, 398–401
Abortion: ancient beliefs, 530–31; laws continually contested, 426; *Roe v. Wade,* 423
Ache (South America), scarification, 469
Adolescent Family Life Act (The Chastity Act), 1982, message of abstinence, 518
Africa: earlobe stretching in traditional societies, 101–2; history of female genital cutting, 227–28; lip stretching, 330, 331, 334–35; neck rings, 355–56
African American rhinoplasty, 375–76
Africanus, Linnaean system, 460
Afro, hairstyle, 249–50, 262–63
Aging, skin, 455–57
Airy, George Biddell, first spherical lens for astigmatism, 124
Alaskan peninsula, ear piercing, 94
Aleutian tribes: ear piercing, 94; lip piercing, 447; nose piercing, 367
Allan, E. G., limb-lengthening, 322
Allor, Dustin, split tongue, 514–15

Amazon River region tribes, lip enlargement, 331–32
American Academy of Cosmetic Surgery (AACS), blepharoplasty, 106
American Academy of Facial Plastic and Reconstructive Surgery (AAFPRS), blepharoplasty, 106
American Academy of Orthopaedic Surgeons, foot or ankle problems, 197
American Birth Control League and Planned Parenthood, Margaret Sanger, 419, 419*f*
American Journal of Obstetrics and Gynecology, standards for obstetrics, 430
American Podiatric Medical Association, foot problems, 197
American Society for Psychoprophylaxis in Obstetrics (ASPO), 431
American Society of Plastic Surgeons (ASPS): antiaging surgery, 149; blepharoplasty, 106, 149; chin augmentations, 77; otoplasty, 103; pectoral implants, 74–75; tummy tuck, 1
Americanus, Linnaean system, 460
Amputation, legs, 317–18
Andronovo tradition, earrings, 95
Anesthesia, use in childbirth, 430–31
Annular vault modification, 288
Anorexia nervosa, 179–80
Anthony, Susan B., anti-corset movement, 544

Antiaging treatments: body, 152–53; cosmetics and beauty creams, 138–39; fat injections, 189–91; feminist views, 158–59; hair, 155; health and illness, 157; history, 137–41; medical procedures, 148–59; medicine and science, 139–41; myths, stories, and quests, 137–38; neck, 152; nonsurgical, 153–55; reality television and, 156–57; surgery, 150–53. *See also* specific surgery
Anti-carbohydrate diets, 174
Antidepressants, first use in 1950s, 19
Anus, taboos and sexual pleasures, 61–62
Apadravya (ampallang and foreskin piercing), 231–32
Apadydoe piercing, 234
Apocalypse Now, mouth symbolism, 338
April Ashley's Odyssey, autobiography of transsexuality, 244
Arden, Elizabeth (Florence Nightingale Graham), beauty industry, 145–46
Aristotle: cardiocentrism, 300; concepts of soul and mind, 12; epigenesis, beginnings of life, 380; on female orgasm, 526; female physical nature, 529–30; inseparability of soul and body, 14; reference to oviducts, 379

Arms, limb-lengthening surgery, 321–24
Art de la Coiffure des Dames (de Rumigny), 251
The Art of Living Long (Cornaro), 139
Asian and Middle Eastern rhinoplasty, 376
Asian Human Rights Commission (AHRC), chastity belt abuse in Rajasthan, 519
Asiaticus, Linnaean system, 460
Assisted Reproductive Technology (ART), 381–82, 413
Astigmatism, 122
AstraZeneca, questionable actions in cancer establishment, 23–24
Atkins Diet, 174
Atlas, Charles, 76
Atlas of Sperm Morphology (Adelman and Cahill), 441
Australia: earlobe stretching, 101; ear piercing, 97
Avar people, earrings, 95

Baartman, Saartjie, Hottentot Venus, 63–64
Baldness, 257
Barbin, Alexina/Abel, nineteenth century intersex individual, 219–20
Bariatric surgery, 177–78, 194–95; body-lift surgery and, 195–96
Beauty ideals: alternative faces, 133–35; anti-racist critiques, 136; cosmetics, 129; feminist critiques, 135; surgery and, 130–33; theory of evolution and, 129; Western, 127–28
Beauty industry: changing trends and conventions, 146–47; debates in the twenty-first century, 147–48; makeup and consumer culture, 145–46
Bedeken, Jewish wedding ritual, 295
Beheading, 286–87
Bellifortis (Kyeser), chastity belt, 517
Benjamin, Harry, transsexuality in the U.S., 238
Best in the World, female bodybuilding competition, 347–48
Bhagavad Gita, images of mouths, 337
Biasutti, Renato, skin color map, 460
Bifocals, 124

Big Beautiful Women (BBW) groups, 188
Binge-eating disorder, 180
Biologically deterministic theory, intersexuality, 221–22
Birth control, 418–28; ancient, folk, and traditional methods, 420; barrier methods, 418–19; behavioral methods, 419; *coitus interruptus,* 420; Depo-Provera, 424–25; emancipation and coercion, 427–28; knowledge and practice, 427; medicalization of, 422–23; morning after pill, FDA approval, 427; 1990s and beyond, 424–25; Norplant implant, 424–25; patch, 425; the pill, 423–24; politics of, 427; voluntary versus compulsory, 422. See also Contraception
Birth control movement, 420–23; late 19th Century, 420
Black booty, 64–65
Bleaching, skin lightening, 463–64
Blepharoplasty, 105–10, 131; antiaging surgery, 151–52; Asian, 107–10, 131; history, 105–7
Blood, **4–11**; cultural history of, 6; cure and anger, 8–11; as a delicacy, 8; humoral theory, 8; magic, Gods, and science, 6–8; menstrual, 7; perceived connection with the soul, 7–8; racial category, WWII, 10; scientific interpretations, 8; symbolic representations, 7, 11
Blood donation, 10–11
Blood groups, 9
Bloodletting, 4–6; conditions, 4; cultural practice, 5; history, 4; leeches, 4–5; sexual practice, 6; specialized equipment, 4–5; spiritual practice, 5
Blood quantum, measurement of racial status, 9–10
Blood sports, 6
Blood transfusions, 8–10
The Bluest Eye (Morrison), hair straightening in African Americans, 262
Bob, hairstyle, 250
Bodybuilding: body image and, 349–50; early years, 346–47; female, 347–48; history, 344–46, 345f; in popular culture, 349

Body Contouring: The New Art of Liposculpture, 513
Body dysmorphic disorder (BDD), cosmetic breast surgery and, 57, 157
Body-lift surgery, 195–96
Body mass index, 168–69
Body modification: branding, 453–54; scarification, 472; subdermal implants, 478–79; tattooing, 488; tongue splitting, 514–15
Body Modification E-zine (Larratt), 451, 503, 514
Body modification movement, 369, 401, 449–52
Body piercing, 446–52; contemporary Western, 449–52; traditional societies, 446–49
Body Play and Modern Primitives Quarterly (Musafar), 451; split tongue, 515
Body-souls, described by Homer, 14
Body weight: gender and fat, 170–71; historical attitudes, 167–68
Body wraps, for cellulite reduction, 165–66
Bondage/domination/sadomasochism (BDSM) culture: chastity belt, 518–19; nose piercing and, 369; play piercing and, 450; surface piercing and, 450
Bonnet, Charles, preformation theory, 380
Book of Optics (Ibn al-Haytham), 115, 123
Botox injections: facial procedure, 131, 153–54; skin procedure, 456
Bourgeois, Louise, the breast in her work, 43
Boys Don't Cry, 1999, representation of transsexuality, 244
Brain, **12–20**; ancient Greeks and the Middle Ages, 13–15; central to mental illness, 18–19; contemporary scientific understanding, 16–17; cultural history, 12–13, 44; imaging, 19–20; mental illness, 17–19; research in the nineteenth century, 18
Branding, 452–54, 472–73; folk surgery procedure, 133–35
Brassiere: emancipation garment, 544; history, 44–45, 45f; politics an Victoria's Secret, 48–49;

precursors to, 46; in women's fashion, 47–48
Brave New World (Huxley), debates around reproductive technologies, 416
Brazil, earlobe stretching, 101
Breast and Cervical Cancer Prevention and Treatment Act of 2000, 23
Breast cancer, 21–25; demographics and diagnosis, 21–22; environmental causes, 23; history, politics, and rise of a social movement, 22–25
Breast Cancer Action (BCA), 23
Breast Cancer Awareness Month, controversy on ICI, 24
Breast cancer genes (BRCA1 and BRCA2), 22
Breastfeeding, 25–29, 26f; versus bottle-feeding, 39–40; colonialism and, 40–41; earliest historical details, 34–35; between nature and culture, 33–34; race and class, 27–28; religion and, 25–27; technology, industry, and 20th Century social movements, 28–29
Breast ironing, 29–31; objections, 30–31; prevalence and rationale, 29–30; procedure, 30
Breasts, **21–58**; anatomy and physiology, 32; ancient history to Renaissance, 34–35; colonialism and, 40–41; cultural history, 31–44; enlightenment, 38–39; masculinity and, 32–33; nursing Madonna, 35–36; obsession in the U.S., 41–42; reclaiming, 42–43; silicone-enhancement, 50–51, 50f; surgical augmentation, 53–56; surgical reconstruction, 56; surgical reduction, 52–53; vocabulary, 33; where hunger and love meet, 36–37
Breast-themed artifacts, 35
Britain: antiaging treatments, 139; eye makeup in medieval times, 118; makeup use in early modern England, 143–44
Browlift, forehead lift surgery, 151
Brown, Louise, first IVF baby, 413
Buddhism, forked tongues, 514
Bulimia nervosa, 180
Burgen, Arnold, discovery of uses for botulinum toxin, 154

Burou, Georges, vaginoplasty technique in sex reassignment surgery, 239–40
Buttocks, **59–68**; body-lift surgery, 195; cultural history, 59–66; eroticism and aesthetics, 61–62; notable figures, 65; ode to the derrière, 65–66; physiology, 59–61, 60f; racialization and contemporary pop culture, 64–65; surgical reshaping, 66–68

Cake, women's sexuality enterprise, 526
California Breast Cancer Research Program, 1993, 24
Calories Don't Count (Taller), 174
Cameroon, breast ironing, 29–31
Camper, Pieter, theory of facial angle, 285
The Cancer Journals (Lorde), 22, 44
Cardiocentrism, Aristotle, 14, 300
Caribbean, ear piercing, 97
Castration, 502–7; medicine and sexuality, 502–3; myth and religion, 504–6
Catlin, George, study of O-Kee-Pa, 72
Caucasian Bronze Age, ear piercing, 95
Cauldwell, David O., advice column in *Sexology* magazine, 238
Cell theory, beginning of life, 380–81
Cellulite: definition, 165; reduction, 165–67
Centers for Disease Control and Prevention (CDC), statistics on birth control use, 425
Cesarean sections, childbirth, 533
Chamberlen, Peter, development of obstetrical forceps, 429
Character: association with the human head, 284–85; hands as a reflection of, 270
Charles, London (Deelishis), buttocks in pop culture, 65
Chastity belt, 517–19, 518f; anti-rape device in South Africa, 519; contemporary, 518–19
Cheeks, **69–71**; fat transfer, 70; implants, 70; surgical reshaping, 69
Chemical peel, facial skin treatment, 131–32, 154–55

Chest, **72–76**; pectoral implants, 74–76
Chibcha tradition: ear piercing, 97; earlobe stretching, 101
Childbirth: anesthesia, 430–31; Cesarean section, 533; epidural analgesia, 433; history in the U.S., 428–32; Kegel exercise, 534; Lamaze technique, 431–32; lithotomy position, 533; natural, 431–32; psychoanalytic theory, 535; sensual birth, 534; standards for obstetrics, 430; vagina in, 532–35. *See also* Pregnancy
Child Rights Information Network, breast ironing, 30
Chin, **77–79**; beauty ideals, 77–78; implants, 78–79; mentoplasty, 78; microgenia, 78; patient profiles for mentoplasty, 78; surgical reshaping, 77
China: castration, 505; ear piercing, 446
Chiricahuas, ear piercing, 96
Chlorpromazine, treatment of schizophrenia, 19
Chopines, type of shoe in medieval Europe, 213
Christian iconography: depiction of the body, 346; hands in, 271–72
Christianity: birth control and, 420; blood of Jesus, 7; castration, 504–5; circumcision and, 393–94; forked tongues, 514; penis redefined by, 386; veiling, 295–96
Christina piercing, 234
Circumcision, 384, 390–96; controversy, 395; cultural, 393–95; medical reasons, 392; methods, 390f, 391; physiology, 391
Clitoridectomy, 226–27, 524
Clitoris, **80–85**; contemporary films featuring, 84; cultural history, 80–85, 522; definition, 80; female genital cutting, 225; piercings, 84, 233; in sex reassignment surgery, 240–41
Clogs, Middle Ages in Europe, 213
Code of Hammurabi, breastfeeding, 25
Collagen: injections, 456; lip enhancement and augmentation, 326–27, 332; skin protein, 455; stretch marks, 476–77

Collagen induction therapy (CIT), facial antiaging treatment, 155
Columbo, Renaldus, discoverer of clitoris, 522
Communication, with hands, 266–68
Comstock Laws, 1873: birth control and, 421; Sanger and, 421
Consciousness, mouth symbolism, 339
Constructionist theory, intersexuality, 1960s, 221
Contact lenses, 126
Contraceptives: access and constraint, 425–28; developments and controversies, 423–24; hormonal, 381, 418; long-term applications, 424–25. *See also* Birth control
Contrappasto, legs in fine art, 316
Conundrum (Morris) 1974, memoir of transsexuality, 244
Cook, Captain James, tattooing, 484
The Correction of Featural Imperfections (Miller), 351
Corset: criticisms of, 543–44; feminist response, 544; as a fetish, 544; first appearance and early versions, 540–43, 541*f*; Gottlieb Oelssner's criticism, 542, 544; history of, 540–45; Isadora Duncan's criticism, 543; Jean-Baptiste Winslow criticism, 542; John Locke's criticism, 542
Cosmetic dentistry, 343, 493–97; modern, 495–96; relevance in contemporary society, 496–97
Cosmetic surgery: abdominal etching, 193; antiaging, 149–50, 149*f*, 156; Asian blepharoplasty, 107–10; blepharoplasty, 105–10; breast, 52–56; cheekbones, 69–71; designer vagina, 525, 538–39; for facial beauty, 130–33; fat injections, 189–91; feminist views, 158–59; health and illness, 157; intranasal rhinoplasty, 373; jaw, 308–10; labiaplasty (labioplasty), 311–13; lip enhancement, 326–27, 332; lip lift, 333; liposuction. *See* Liposuction; mentoplasty, 77–78; neck lift, 351–53; Orlan, 162; otoplasty, 103–4; pectoral implants, 74–76; psychology of, 57; reality television and, 156–57; rhinoplasty, 363–64, 370–78. *See also* Surgery, and the specific procedure
Cosmetics: antiaging, 138–39; changing trends and conventions, 146–47; facial beauty and, 129–30. *See also* Makeup
Council on Size and Weight Discrimination, 185
Cradle boarding, 289
Crakows, type of shoe in medieval Europe, 213
Cranial binding, 289
Cranial modification. *See* Head shaping
Craniometry, 285–86
The Crying Game, 1992, representation of transsexuality, 244
Cultural practices: anthropological understandings of the mouth, 339–40; blood and, 6–8; bloodletting, 5; bodybuilding, 344–50; feet, 200–201; footbinding, 206–7; lipstick, 326; menopause, 405–6; nose, 360–66; privileges of lighter skin, 474–75; rhinoplasty, 376–77; significance of the heart, 302

The Daily Mirror Beauty Book, 1910, eye makeup, 119–20
Dances Sacred and Profane, flesh-hook suspension, 450
Dandy, Walter, brain imaging, 19
da Vinci, Leonardo, penis in public art, 386
Dawenkou, ancient earrings, 93–94
Dear Sir or Madam: The Autobiography of Mark Rees, 1996, 244
Death, mouth association with, 337–38
Decade of the brain. 1990s, President George H. W. Bush, 20
Decorated Skin: A World Survey of Body Art (Gröning), nose ornamentation, 368
De Curtorum Chirurgia (Tagliacozzi), first rhinoplasty, 372
Deep hood piercing, 233
de Gardanne, C. P. L., first to identify menopause, 403
Degeneration theory, hereditary mental illness, 18
de Grey, Aubrey, aging definition, 140
De humani corpus fabrica (On the Fabric of the Human Body) (Vesalius), 386
Dental transfiguration, 493–95; chipping and filling, 494–95; extraction, 494
Dentistry, depictions in contemporary art, 342–43
Dentures, 494
Department of Defense Breast Cancer Research Program, 1993, 24
De re anatomica (Colombo), discovery of clitoris, 80
Dermabrasion, facial skin treatment, 131–32, 154–55, 457
Dermatology, 454–55; origins of skin lightening, 473–74
Dermis, inner layer of skin, 455
de Rumigny, Legros, education of hairdressers, 251
Descartes, René, mind-body problem, 12, 14–16
Designer vagina, 525, 538–39
Desperate Living, 1979, transsexual film, 244
de Tours, J. Moreau, degeneration theory, 18
Deviant Bodies: Critical Perspectives on Difference in Science and Popular Culture (Terry & Urla), 161
Diagnostic and Statistical Manual for Mental Disorders (DSM-IV): borderline personality disorder (BPD), 464–65; eating disorders, 179; gender identity disorder (GID), 236
Diamond, Milton, research into intersexuality, 221–22
Dickinson, Robert Latou, sperm and semen, 440
Dieffenbach, Johann: nasal surgery, 364; post-traumatic correction of the ear, 103; rhinoplasty, 372
Diet and Health, With Key to the Calories (Peters), first low-calorie diet book, 173
Diet Breakers (England), fat activism, 184
Diet pills, 174
Dieting practices, 172–78; bariatric surgery, 177–78; cigarettes

and, 173; control, and loss of pleasure, 176–77; criticisms, 175–77; diet pills, 174; fad dieting, 173–75; obesity as a global epidemic, 177; origins, 173; weight-loss companies, 175
Dinitrophenol, early diet drug, 174
Diocles of Carystus, discovery of the ovaries, 379
Disability rights movement, resistance to limb-lengthening surgery, 323
Dog Day Afternoon, 1975, representation of transsexuality, 244
Dogon (Africa), piercing, 448
Dominican Republic, occurrence of intersex condition, 218
Donor insemination (DI), 413
Donor siblings, conceived from same sperm donor, 438*f*
Dr. Charles Conrad Miller's Review of Plastic and Reconstructive Surgery, eye surgery, 106
Dumas, Jean-Baptiste, scientists view of sperm and fertilization, 440
Duncan, Isadora, against restrictive style of dress, 543
Durston, William, breast removal, 52
Dwarfism, limb-lengthening surgery, 318, 321–24, 322*f*
Dydoe piercing, 234
Dysport injections, facial procedure, 131

Ear, **86–104**; adornment, 89; cropping and shaping, 91–93; cultural history, 86–91; cutting off as punishment, 89; earlobe stretching, 100–102; human-mouse ears, 88; modifications, 87–88; otoplasty, 88, 102–4; physiology, 86–87; piercing, 89, 93–99, 93*f*; symbolic representations, 89–91
Eating: anthropological understandings of the mouth, 339–40; normal versus disordered, 179
Eating disorders, 178–83; demographics, 181; emerging research, 181–82; etiology, 180–81. *See also* specific disorder
Egypt (ancient): bathing of feet, 200; birth control, 420; dentistry, 494; ear piercing, 94; earlobe stretching, 100; eye in religion, 111–12; eye makeup in antiquity, 117; hair removal, 258; heart ritual, 298–99; lip enlargement, 331; makeup use, 142–43; Mehndi, 274; penis as phallus, 385; pubic hair removal, 264; skin care, 456
Elastin, skin protein, 455
Elbe, Lili, transsexual, 243
Ely, Edward Talbot, cosmetic otoplasty, 103
Emancipation garments, 46–47
Emergence: An Autobiography (Martino) 1977, 244
Émile (Rousseau), enlightenment on the breast, 38
Emotion, cultural meaning of the heart, 302–3
Encephalocentric theory, 14
Endermologie, for cellulite reduction, 166
Endoscopic browlift surgery, 151
Ensler, Eve, *The Vagina Monologues,* 536–37, 536*f*
Epic of Gilgamesh, images of mouths, 337
Epidermis, outer layer of skin, 455
Epidural analgesia, use in childbirth, 433
Epigenesis theory, 380
Episiotomy, 533
Eschappins (escarfignons), type of shoe in medieval Europe, 213
Eskimos, nose piercing, 367
Estrogen: birth control, 418; during menopause, 402–3
Etruscans, ear piercing, 95
Eunuchs, in antiquity, 507–11
Europe: eye makeup in medieval times, 118–19; leg adornment, 16th and 17th Century, 314
European Society of Human Reproduction and Embryology (ESHRE), 414
Europeanus, Linnaean system, 460
Evil eye mythology, 110–11
Evolution, explanations of beauty and, 129
Extended limb-lengthening surgery (ELL), 318, 321–24, 322*f*
Eyeglass making guilds, fifteenth and sixteenth centuries, 124
Eyelid surgery. *See* Blepharoplasty
Eye makeup: antiquity, 116–18; changing trends in the modern era, 119–22; history, 116–22; middle ages to modernity, 118–19
Eye of Horus, 111–12
Eye of Providence, 112
Eyes, **105–26**; cultural history, 110–16; evolution or intelligent design?, 115–16; makeup. *See also* Eye makeup
Eyewear, history, 122–26

Fabrey, William, founder of NAAFA, 183
Face, **127–64**; alternatives, folk surgery, 133–35; beauty surgery, 130–33; cosmetics, 129–30; cultural ideals of beauty, 127–28; facial beauty and the theory of evolution, 129; performance art, 161–64; teeth and gums in beauty ideal, 132–33
Facial implants, 132
Facial lips, biology of, 325–26
Fad dieting, 173–75
False eyelashes, 120–22
Family lineage, blood and, 9
Family planning, voluntary, 426–27
Fashion: eyewear as an accouterment, 125; hair removal and, 259; hair removal from legs, 315–16; hats and head adornment, 282; legs, 314–15; lipstick, 326; practical applications of the bra, 47–48
Fat, **165–96**; activists and pride, 171–72, 183–88, 183*f*; cellulite reduction, 165–67; cultural history, 167–72; eating disorders, 178–83; injections, 189–91; Overeaters Anonymous, 171. *See also* Body weight; Dieting practices; Obesity
Fat activism, 171–72, 183–88, 183*f*; connections to other social movements, 187–88; criticism, 188; current key organizations and leaders, 184–86; health and fat, 186–87; social organizations, 188; terminology, 186
Fat and Proud (Cooper), 185
Fat is a Feminist Issue (Orbach), 176

Fat Power: Whatever You Weigh is Right (Louderback), 187
Fat transfer, 189–91; lip enhancement procedure, 327, 332
Fat Underground, 184
Fat Women's Group (England), fat activism, 184
Fat?So! (Wann), 185
Feet, **197–215**; absence of, athletes, 190; art and literature, 199; cultural and religious customs, 200–201; cultural history, 197–201; military and athletics, 197–98; sexuality and fetishes, 199–200; unusual and famous, 198–99
Female genital cutting (FGC), 83, 224–25*f*, 224–31, 523–24; Africa and the Islamic world, 227–28; critiques, 228; eradication efforts, past and present, 230; history in Western medicine, 226–27; overview, 224–25; proponents, 228–30; types, 225–26; WHO definition, 225
Female physical nature, ancient beliefs, 529–30
Feminine Forever (Wilson), menopause, 405
Feminism: anti-corset movement, 544; birth control, 419; birth control movement, 419–21; bra protest, 45, 45*f*, 48–49; chastity belt, 517–18; Comstock Laws and, 420; cosmetic surgery and antiaging treatments, 158–59; cosmetic surgery critique, 54–55; critiques of beauty ideals, 135; fat-activist movement and, 184, 187; fat and the female body, 176–77; fertility treatments and, 416–17; the gaze, 112–13; high-heeled shoes, 200, 214; menopause and, 405; Orlan, 161–64, 162*f*; penis envy and, 396–98; reclaiming the breast, 42–43; sex reassignment and, 241; sexual empowerment and freedom, 526; vaginal rejuvenation concerns, 538–39; veiling and, 294–95
Feminist International Network of Resistance to Reproductive and Genetic Engineering (FINRRAGE), 417

Feminist Mass Meeting, 1914, right to ignore fashion, 135
Femur (thighbone), use in making trumpet called Rkang Dung, 316
Ferrari, Lola, largest silicone-enhanced breasts, 50–51, 50*f*
Fertility technologies: contemporary biomedicine, 413; global issues, 417–18; harvesting and freezing eggs, 382; oppositional voices, 414–17; ovaries and, 381–82; premodern treatments, 412–13; sperm research, 440–41; treatments, 411–18
Fetish: buttocks, 61–62; chastity belt, 518–19; corset, 544; feet, 199–200; shoes, 214; thighs, 513
Fiji, subincision, 400
The First Wives Club, inflated lips, 331
Fischer, Giorgio, liposuction, 512
Flat feet (pes planus or fallen arches), 198
Flesh-hook pulling: Fakir Musafar, 450; Hindus of India, 449; Mandan O-Kee-Pa ceremony, 449; Thaipusam festival, 448
Flourens, Pierre, holistic understanding of brain function, 17
fMRI (functional magnetic resonance imaging), brain imaging, 19–20
Folk surgery, 133–35
Foot binding, 200, 201–8, 202*f*; binding process, 204–5; cultural practices, 206–7, 211–12; decline of, 207–8; origin, 202–3; sexual rituals, 206; shoelessness and, 211–12; symbolism, 205–6
Forehead lift, facial surgery, 131
Forked tongues, 514
Frankenstein technologies, reproductive technologies media phrase, 416
Freeman, Walter, lobotomy in the U.S., 18–19
Freemasons, eye of Providence, 112
Free will concept, 12
Frenum ladder, 234
Frenum piercing, 234
Freud, Sigmund: anus and buttocks, 61–62; castration, 503; clitoris, 522–23; Oedipal view of vagina, 535; penis as symbol, 388; penis envy, 396–98
Friedman, Emanuel, obstetrical model of labor, 433
Friedman's curve, model of labor, 433

Ga'anda (Nigeria): ear piercing, 97, 448; scarification, 468–69
Galen: ovaries, 379; physical basis of mental illness, 17
Galilei, Galileo, optical modification, 125
Gall, Franz Joseph: location of the mind, 16–17; phrenology, 285
Gastric banding, 177, 194–95
Gastric bypass surgery, 177, 191*f*, 194–95
The Gaze, cultural significance, 112–13
Gbinna (Niger-Congo), scarification, 469
Gender: bloodletting, 5; breast ironing, 30–31; dieting and, 175–76; eroticism of the buttocks, 61–62; facial beauty ideals, 132; fatness and, 170–71; hairstyle changes, 250; jaw ideals of beauty, 308; penis envy, 396–98; reassignment surgery. *See* Sex Reassignment surgery; veiling, 292–97
The Gender Frontier (Allen), gender variant photography, 244
Gender identity disorder (GID), 236
Genitals, **216–45**; cultural history of intersexuality, 216–23; female genital cutting, 224–25*f*, 224–31; piercing, 231–35; sex reassignment surgery, 235–44; *The Vagina Monologues,* 536–37
Gersuny, Robert, breast augmentation, 54
Gillies, Harold: pioneer of phalloplasty, 241; rhinoplasty techniques, 373
Glen or Glenda (I Changed My Sex) (film), version of Christine Jorgensen story, 244
Global Strategy on Diet, Physical Activity and Health (WHO), 169
Gluteus maximus, 59
Gomco clamp, use for circumcision, 391

Gore-Tex implants: lip enhancement, 327; nasal, 374
Great Competition, bodybuilding, 1901, 347
Greece (ancient): antiaging, 138; birth control, 420; bodybuilding, 344–46; castration, 504; ear piercing, 95; eunuchs, 508–9; eye makeup in antiquity, 117; hair removal, 258; heart importance, 299–301; hysteria in women, 521–22; idea of facial beauty, 127–28; ideal nose, 361–62; legwear, 315; makeup use, 143; penis as phallus, 385; pubic hair removal, 264; sexual intercourse and women's health, 526; skin care, 456; vagina and childbirth, 532
Greek matrimony, 278–79; the Kyrios (father), 279; the wedding, 278–79
Griesinger, Wilhelm, brain research, 18
Griswold v. Connecticut, contraception legalized throughout the U.S., 423–24
Guiche piercing, 234
Guinand, Pierre-Louis, production of optical glass, 124
Gynecologists, 19th Century, 380–81
Gynecomastia, 32–33

Hadfield, James, insanity defense, 18
Hafada piercing, 234
Hafner's frame fitting chart, 1898, 122*f*
Hair, **246–65**; color, 252–53; cultural history, 246–55; economics of, 250–51; modifications, 251–52; physiology, 246–47; politics of, 249–50; symbolic meanings, 247–49
Hair removal, 256–60; body, 258–60; head, 256–58; legs, 315–16; methods, 258
Hair replacement, antiaging treatments, 155
Hair straightening, 253–54, 260–63; methods, 261–62, 261*f*; racial politics, 262–63
Hall, Thomas/Thomasine, first case of intersex individual, 218
Hamadsha sect, head slashing, 290–92

Handedness, 272
Handfasting, 279–80
Hands, **266–80**; character and spirituality, 270–72; communication and expression, 266–68; cultural history, 266–74; forms of deception with, 269–70; gang tattoos, 268; the hand in marriage, 276–80; methods of adornment, 267; pagan and medieval handfasting, 279–80; training and work, 268–70
Harmony, in facial beauty, 127–28
Hartsoeker, Nicholas, theory on sperm, 439–40
Harvey, William, seventeenth century study of the heart, 301
Hats: head protection, 283; indication of inclusion or solidarity, 282–83; since antiquity, 282
Haworth, Steve, body-modification pioneer, 478–79
Hay Diet, 174
Head, **281–97**; adornment and display, 281–84; cultural history, 281–88; intellect, character, and personality, 284–86; punishment and violence, 286–87; veiling, 292–97, 292*f*
Headhunting, 287
Head shaping, 288–90; contemporary, Western medicine, 289–90; meaning, 289; methods, 288–89
Head slashing, 290–92
Health and Strength Magazine, 1898, bodybuilding, 346–47
Health at Every Size Movement, 187
Hearing loss, 87
Heart, **298–305**; cultural history, 298–305; cultural significance, 302; early beliefs, 298–301; emotion and, 302–3; Renaissance period, 301–2; spiritual significance, 303–4; symbol, 299*f,* 304–5
Heart of Darkness (Conrad), mouth symbolism, 338
Height-weight tables, 168–69
Heinroth, J. C. A., brain research, 18
Helmets, head protection and decoration, 283–84
Henna, use in Mehndi, 267, 275
Hermaphrodite. *See* Intersexuality

Herophilus, discovery of the ovaries, 379
Herschel, Sir John, contact lenses, 126
Hertwig, Oscar, beginning of life, 380–81
Hijab, 292*f,* 293–94
Hippocrates: female physical nature, 529–30; hysteria in women, 521–22; on menopause, 404; physical basis of mental illness, 17; sexual intercourse and women's health, 526–27
Hirschfeld, Magnus, study of gender-variant people, 238
Hitler, Adolph, idea of racial blood, 10
HIV: circumcision effects on transmission, 393; female genital cutting and, 228
Hollander, Eugen, first antiaging cosmetic surgery, 150
Hollywood diet (grapefruit), 174
Holocaust, tattooing of prisoners, 485
Homer, two different kinds of souls, 14
Hongshan, ancient earrings, 93
Horiyoshi, Tamotsu, tattoo artist, 480*f*
Hormone replacement therapy (HRT), use in menopause, 405
Hottentot Venus, 63–64
Human Fertilisation and Embryology Authority (HFEA), U.K., 114
Human papillomavirus (HPV), circumcision effects on transmission, 392
Human Sex Anatomy (Dickinson), 440
Human Sexual Response (Masters & Johnson), 523
Human Sperm Competition: Copulation, Masturbation, and Infidelity (Baker and Bellis), 442–43
Humoral theory of blood, 8
Hymen, **306–7**, 521; surgical restoration, 306–7
Hymenoplasty, 306–7
Hysteria: female physical nature, 529–30; medieval physicians, 521–22; 19th and 29th Centuries, 531; symptoms, 527–28; treatment, 528–29; wandering womb and, 526–31

Ilizarov, G., innovations in limb-lengthening, 1951
Illouz, Yves-Gerard, modern liposuction, 512
Imperial Chemicals Industry (ICI), controversy on Breast Cancer Awareness Month, 24
Impotence, 388
India: ear piercing, 96; earlobe stretching, 101; Eunuch's conference, 217*f*; eye makeup in antiquity, 117–18; first cosmetic surgery, 130; flesh-hook pulling, 449; hijras (gender variant people), 242; hijras and sadhins (intersexual individuals), 216–17; nose piercing, 368; significance of feet, 201; surrogate mothers, 412*f*
Infant formulas, 28
Infibulation, 226; male, 232–33
Injectable dermal filters, facial antiaging treatment, 153
Insanity defense, Hadfield case, 18
Institute for Sexual Research (Berlin), 238
Institute of Physical Culture, 1897 London, Eugen Sandow, 345*f*, 346–47
Integumentary system, biology of skin, 455
Intellect, association with the human head, 284–85
International Size Acceptance Association, 185
Intersex Society of North America, 223
Intersexuality: contemporary treatment, 222–23; cross-cultural treatment, 216–18; definition, 216; medicalization of, 219–22; Western treatment before medicalization, 218–19
Intracytoplasmic sperm injection (ICSI), 414
Intrauterine insemination (IUI), 413
In vitro fertilization (IVF), 413
Iraq, nose piercing, 446
Irezumi, tattooing in Japan, 481–82
Irish rhinoplasty, 375
Isabella piercing, 233
Islam: castration, 505; circumcision and, 394; history of female genital cutting, 227–28
Italy, first corrective eyeglasses, 123

Jackson, Janet, buttocks in pop culture, 65
Jackson, Michael: face, 159–61, 159*f*; nose, 366, 376–77; vitiligo (skin lightening), 474
Jaw, **308–10**; surgical reshaping, 308–10
Jeans, invention in 1873 by Levi Strauss, 315
Jefferson skeletal classification, facial beauty, 128
Jeffreys, Sheila, beauty practices, 135
Jenny Craig diet program, 175
Jewish Rhinoplasty, 375
Johns Hopkins University Medical Center, sex reassignment surgery, 239
Jorgensen, Christine (George), first public sex change, 1950, 238
Joseph, Jacques, breast plastic surgery, 52
Judaism: castration, 504; circumcision and, 393; veiling, 295

Kama Sutra: genital piercing, 231–32; male genital piercing, 447; poses to increase sexual pleasure, 525
Kamikaze Sperm Hypothesis (KSH), 442–43
Kant, Immanuel, location of the soul, 16
Karen tribe (Myanmar, Burma), neck rings, 353–56, 353*f*
Kayapo (Brazil), lip stretching, 335
Kegel, Arnold, exercise for childbirth, 534
Kelly, H. A., first abdominoplasty, 1
Kepler, Johannes, optical modification, 125
Khoi Khoi, Hottentot venus, 63
Khoo Boo-Chai, modern Asian blepharoplasty, 108
Kito, Shimo, Asian blepharoplasty, 108
Kuna (Panama): ear piercing, 97; nose piercing, 368–69
Kushner, Rose, breast cancer activist, 22
Kyeser, Conrad, first presented chastity belt, 517

Labia, **311–13**; piercing, 233
Labiaplasty (labioplasty), 311–13
Labret piercing, lips, 330
Lactation, 32
La Leche League, supporting breastfeeding, 40

Lamaze technique (psychoprophylaxis), childbirth, 431–32
La Réincarnation de Sainte Orlan (The Reincarnation of St. Orlan), 162–64, 478
LASER, antiaging surgery, 152
Laser resurfacing, facial skin treatment, 131–32
LASIK eye surgery, 126
Late Andean Formative tradition, nose piercing, 367
Late Hohokam tradition, nose piercing, 368, 447
Leadership, use of word head, 284
Leeches, used in bloodletting, 4–5
Legs, **314–20**; adornment and clothing, 314–15; amputation, 317–18; cultural history, 314–20; hair removal, 315–16; limb-lengthening surgery, 318, 321–24, 322*f*; literature, 319–20; music and fine art, 316–17; myth, legend and religion, 318–19
Leighton, Frederic Lord, *The Sluggard,* depiction of male form, 76
A Letter on Corpulence Addressed to the Public (Banting), 173
Lex de Maritandis Ordinibus, 277
Lillie, Frank, scientists view of sperm and fertilization, 440
Lima, Almeida, first lobotomy, 18
Limbs, **321–24**
Linnaeus, Carolus, classification/taxonomic system for humans, 460
Lip lift, 333
Liposuction: abdominal, 2*f*, 3, 152–53, 192; neck, 351; thigh, 512–13
Lips, **325–36**; biology of, 325–26; cultural history, 325–31; enhancement and augmentation, 326–27; enlargement, 331–33; implants, 332–33; piercing and stretching, 329–31; racialization and, 327–29; stretching, 333–36, 334*f*
Lipstick, 326
Listening to Mothers (Maternity Center Association), 433
Lithotomy position, childbirth, 533
Lobotomy, 18–19
Locke, John, crusade against the corset, 542

Loeb, Jacques, scientists view of sperm and fertilization, 440
Lombrosos, Cesare: craniometry, 285–86; evolution and criminal pathology, 18
Loomis, Roland, recreation of O-Kee-Pa, 73–74
Lopez, Jennifer, buttocks in pop culture, 65
Lorde, Audre, breast cancer activist, 22
Lorgnettes, 125–26
Lorum piercing, 234
Lotus feet, Japan, 202, 202f
Louis XV of France (1710–1774), hairdressing takes form, 250
Love, Susan, breast cancer activist, 23

Maasai of Matapato, earlobe stretching and cutting, 102
Madame de Pompadour (1721–1764), hairstyles, 250
Madonna and Child (Madonna Litta) (da Vinci), 35–36
Madonna's breast, 35–36
Magnoan, Valentin, evolution and criminal pathology, 18
Majabang, ancient earrings, 93
Makeup: aging skin, 457; beauty industry and consumer culture, 145–46; changing trends and conventions, 146–47; debates in the twenty-first century, 147–48; history, 141–48; middle ages and early modern England, 143–44; twentieth century to present, 144–45; use in antiquity, 141f, 142–43. See also Cosmetics; Eye makeup
Malarplasty: augmentation, 69–70; reduction, 70–71
Male breasts, 32
Malloy, Doug, innovator in body piercing, 450
Malone, Annie Turnbo, beauty industry, 145–46
Mammography, 22
Mandan tribe, O-Kee-Pa, 72–74, 449
Man Into Woman (Elbe), 243
Manual labor, working with ones hands, 269
Maori (New Zealand), tattooing, 481–82
Marquardt Beauty Analysis (MBA), facial beauty, 128

Marriage, hand in marriage, 276–80
Mascara, 120–22
Masculinity, pectoral implants, 74–76
Mastectomy (Halsted), 22; reconstruction after, 56
Masturbation, 386–87
Matlock, David, designer vagina surgeon, 538
Mayan culture, bloodletting in, 5
Meddling with nature, reproductive technologies, 415–16
Medusa myth, evil eye myth, 111
Mehndi: designs, 275–76; India Day Festival, 274f; method of adorning hands, 267; origins, 274–75
Melanin, skin color, 457
Menopause: cultural history, 402–7; culture and, 405–6; definition, 402; medicine and, 403–5; physiologic and psychologic changes, 403; in popular culture, 407
Men's fertility, 440–41
Menstrual blood, cultural beliefs, 7
Menstrual pads, 434–36
Menstruation: cultural history, 407–11; practices and products, 434–36; predecessors to tampons and pads, 435; taboos, 409–11
Mental illness, 17–19; brain as central explanation, 18–19; confinement in large institutions, 17–18; Greek art and theater, 17; hereditary, 18; Middle Ages in Europe, 17; nineteenth century, 18; skin cutting, 464–66
Mentoplasty, 77–78, 309–10; patient profiles, 78; procedures, 78–79
Mesolithic Stone age, trephination, 490
Mesotherapy, for cellulite reduction, 166
Metchnikoff, Elie, cell degeneration and aging, 140
Metzger, Deena, breast cancer activist, 22
Mexican tradition, nose piercing, 367
Mexico: ear piercing, 97; scarification, 467

Michael Jackson's Face (British-made documentary), 159–61, 159f
Microdermabrasion, facial skin treatment, 131–32, 457
Microgenia, 78
Microscope: enhancement of vision, 113; scientists view of sperm, 439
Middle Ages: bloodletting in, 5; madness in, 17; makeup use, 143
Middle East, ancient times ear piercing, 94
Midwifery, 428–32; sensual childbirth, 534; versus physician-attended birth, 429–30
Mikamo, K., first Asian blepharoplasty, 107
Millard, D. Ralph, Asian blepharoplasty, 108–9
Miller, Charles C., aesthetic eye surgery, 105–6; development of neck lift techniques, 351
Mind, concept of, 12
Mind-body problem, Descartes, René, 12, 14–16
Minoan culture: ear piercing, 94–95; predecessor to the corset, 540–41
Modern primitives: body modification movement, 449–52; nose piercing and, 369; scarification, 472; subincision, 401; tattooing, 488
Modern Primitives, 450
Moench, G. L., sperm morphology, 441
The Monastery (Scott), handfasting ceremony, 279–80
Money, John, psychology and the intersex individual, 1960s, 221
Moniz, António Egas, first lobotomy, 18
Monocles, 125–26
Monroe piercing, lips, 330
Morel, Benedict Augustin, degeneration theory, 18
Morestin, Hippolyte, rhinoplasty techniques, 373
Morton, Samuel, cranial capacity theory, 285
Mouth, **337–43**; anthropological understandings, 339–40; cultural history, 337–43; symbolism, 337; teeth, 340–42

Mr. Universe competition, bodybuilding, 1948, 347
Mulvey, Laura, feminist film theorist, 112
Mursi group, Ethiopia, lip stretching, 330, 331, 335
Musafar, Fakir: flesh-hook suspension in the U.S., 450; scarification, 472; tattooing, 488
Muscle dysmorphia, bodybuilding and, 349–50
Muscles, **344–50**; cultural history of bodybuilding, 344–50; female bodybuilding, 347–48; steroids, 348–49
Muslim cultures, pubic hair removal, 264
Myopia (nearsightedness), 122
Myra Breckinridge, 1970, transsexual film, 244
Mythology, eyes in, 110–12

Nasal index, 362–63
Nasallang, 370
Nasology, 328, 363
National Alliance for Breastfeeding Advocacy, 28
National Amateur Bodybuilders Association (NABBA), Miss Universe competition, 1965, 347
National Association to Advance Fat Acceptance (NAAFA), 172, 183–87, 183*f*
National Breast and Cervical Cancer Early Detection Program, 1990, 23
National Breast Cancer Coalition (NBCC), research, screening, and care, 23
National Childbirth Trust, supporting breastfeeding, 40
National Fat Women's Conference, 1989, 184
National Organization of Lesbians of Size (NOLOSE), 185
National Organization of Women, opposes size discrimination, 184
Native Americans: hair removal, 247–48, 258; lip stretching, 335; nadles and berdaches (intersexual individuals), 216–17; piercing and flesh-hook pulling, 448–49; piercings, 96
Natural Childbirth (Dick-Read), 431
Natural childbirth movement, 431–32

Natural History (Pliny the Elder), 138
Ndebele tribe (South Africa), neck rings, 355–56
Neck, **351–56**; lift, 351–53
Neck rings, 353–56, 353*f*
Nefertiti piercing, 233
New Haven Fat Liberation Front, 184
New Zealand: ear piercing, 97; earlobe stretching, 101
Nipple, **357–59**; cancer, 359; piercing, 357; removal, 357–59; supernumerary, 358–59
No Diet Day, 184
Normal, 2003, representation of transsexuality, 244
Northwestern University, sex reassignment surgery, 239
Nose, **360–78**; cocaine-damaged, 364; cultural history, 360–66; definition, 361; ideal nose, 361–62; implants, 374; notable, 365–66; piercing, 366–70; rebuilding, 372; rhinoplasty, 363–64, 370–78; rituals and customs, 365; rubbing noses, 361, 361*f*; syphilis-damaged, 364; typing and racialization, 362–63
Nose job. *See* Rhinoplasty
Nuba (Africa), scarification rituals, 468
Nubian Pan Grave culture, earrings, 94

Obesity: causes of and treatments for, 169–70; global epidemic, 169, 177
Obesity Myth (Campo), 185–86
Obstetrical forceps, 429
Obstetricians, 19th Century, 381
Oedipus cycle of Sophocles, mental illness, 17
Oelssner, Gottlieb, criticism of the corset, 542
O'Keefe, Georgia, flower paintings resembling genitalia, 81
O-Kee-Pa, 72–74, 73*f*; piercing and flesh-hook suspension, 449
Oman, xaniths (intersexual individuals), 216
Onanism, term for masturbation, 438
On the Usefulness of the Parts of the Body (Galen), cardiocentrism, 300
Ophthalmology, 114–15

Opposite theory, hair, 248
Orgasm: as cure for hysteria, 522; vaginal, 522–23
Orlan, performance artist, 161–64, 162*f*
Orthognathic procedures, 309–10
Osteotomy, maxillary and/or mandibular, 309
Otoplasty, 88, 102–4; children, 103; human-mouse ears, 88
Ötzi the Iceman, scarification example, 467
Ovarian cancer, 382–83
Ovarian transplants, 383
Ovaries, **379–83**; contraception technologies and, 381; cultural history, 379–83; discovery, 379–81; fertility technologies and, 381–82; harvesting and freezing eggs, 382
Overeaters Anonymous (OA), 171
Oversby, Alan (Mr. Sebastian): genital piercing, 234–35; innovator in body piercing, 450
Ovism theory of preformation, 438–39
Ovists, 17th Century scientists, 380

Padaung tribe (Thailand), neck rings, 353–56, 353*f*
Paget's disease, mammary, 359
Pain, dentistry and, 342–43
Paley, Dror, Ilizarov technique for limb-lengthening, 322
Palmistry, 270–71
Papua New Guinea: earlobe stretching, 101; ear piercing, 97–98; occurrence of intersex condition, 218; scarification, 469
Parasuicide, WHO term, 466
Parker, Geoff, sperm competition, 441–42
Passot, Raymond, first browlift surgery, 151
Paterfamilias, 277
Pater Potestas, 277
Pectoral implants, 74–76
Penis, **384–401**; impotence, 388; invisible and visible, 388–89; male anxieties, 387–88; masturbation, 386–87; as phallus in pagan and non-Western cultures, 385–86; redefined by Christianity, 386; size, 387–88; tattooing and piercing, 389
Penis envy, 396–98

Peropia (farsightedness), 122
Personality, association with the human head, 284–85
Peru: ancient ear ornamentation, 97; earlobe stretching, 101
PET (positron emission tomography), brain imaging, 19–20
Phalloplasty, in sex reassignment surgery, 241
Phi (golden ratio): facial beauty and, 128; ideal chin, 77
Phlebotomy, 4–6
Phrenological symbolic meaning of the human head, 1842, 13*f*
Phrenology, 285, 328, 363; 16–17
Physicians, versus midwives, 429–30
Physiognomy, 308–9, 328
Pianelle, type of shoe in medieval Europe, 213
Piercing: clitoris, 84, 233, 449–52; ears, 89, 93–99, 93*f*, 446–52; facial, 133–35, 447*f*, 449–52; genital, 231–35, 447–52; lips, 329–31, 447–52; nipple, 357, 449–52; nose, 366–70, 446–52; penis, 389, 449–52; scrotal, 234, 449–52. *See also* Body piercing, and specific type or site
Piercing Fans International Quarterly (PFIQ), 450
Pink ribbons, breast cancer symbol, 23–24
Pinna (ear), 86–104
Pistorius, Oscar, athlete with disability, 199
Plastibell, use for circumcision, 391
Plato: concepts of soul and mind, 12; dualism, soul and body division, 14
Platysmaplasty (neck lift), 351–53; antiaging neck surgery, 152
Polydactylism, 198
Polynesia, earlobe stretching, 101
Polytetrafluoroethylene (PTFE): cheek implants, 70; chin implants, 79
Pornography, clitoris, 83–84
Poulaine, type of shoe in medieval Europe, 212
Pousson, Alfred, reduction mammoplasty, 52
The Power of Pleasure (Atrens), food and dieting, 176
Preformation theory, 380; spermism versus ovism, 438–39

Pregnancy: abdominal changes, 2; bloodletting and, 4; contemporary experience, 432–34; harvesting and freezing eggs, 382; prevention: birth control, 418–28; ovaries and, 381; stretch marks, 476–77; termination and prevention, ancient beliefs, 530–31. *See also* Childbirth
Preimplantation genetic diagnosis (PGD), for genetic selection, 416
Presbyopia, 122
Prévost, Pierre, scientists view of sperm and fertilization, 440
Prince Albert piercing, 232, 389
Princess Albertina, 233–34
Pro-life debates, 414–15
Progestin, birth control, 418
The Prolongation of Life (Metchnikoff), 140
Promiscuity: An Evolutionary History of Sperm Competition and Sexual Conflict (Birkhead), 442
Proportion, in facial beauty, 127–28
Prostate cancer, orchiectomy and orchidectomy, 502
Psyche, described by Homer, 14
Psychoprophylactic method, childbirth, 431–32
Psychosexual development, penis envy, 396–98
Psychosexual inversion, 237–38
Psychotropic drugs, new science of brain identity, 19
Puberty: menstruation and, 407; penis and, 385
Pubic hair: removal, 259–60; shaving and waxing, 263–65
Pumping Iron (film), bodybuilding, 349
Pumping Iron II: The Women (film), female bodybuilding, 349
Punishment: branding, 453; castration, 506–7; hair removal as, 256–57; hands and, 272–73; head, 286
Punk culture: nose piercing and, 369; tattooing, 486–87
Putti, Vitorio, Osteoton device for limb-lengthening, 322

Queer movement, tattooing, 489

Race/ethnicity: abdominoplasty rates, 3; the afro, 249–50; blood and measurement of racial status, 9–10; breastfeeding, 27–28; breastfeeding and, 40–41; buttocks and, 62–65; class and skin color, 462–63; cosmetic rhinoplasty and, 374–76; critiques of Western beauty ideals, 136; cultural privileges of lighter skin, 474–75; hair straightening, 262–63; jaw differences, 308–9; lips and, 327–29; nose, 360, 362–63; skin color and, 459–62
Race for the Cure, breast cancer, 23
Radiance, fat-activist feminist magazine, 184
Radiance, lip enhancement, 327
Rebirth, mouth association with, 338
Religion: adornment of the head, 282; eyes in, 110–12; feet in, 200–201; hair removal and, 256; hands in, 271; left- and right-hand path, 272; legs in, 318–19; masturbation, 386–87; symbolism of the heart, 303–4; veiling, 292–97, 292*f*
Renaissance period: anatomic understanding of the heart, 301–2; chastity belt, 517; corset use, 542; muscular male body, 346; penis in public art, 386; women's bodies in art, 2
Reproductive system, **402–36**; disrupting the reproductive order, 415; fertility treatments, 411–18; menopause, 402–7; menstruation, 407–11
Reshaping the Female Body (Davis), 54
Restylane, lip enhancement, 327
Revolutionaries, hairstyles and, 249
Rhinoplasty, 363–64, 370–78, 371*f*; before/after, 377; early, 371–72; implants, 374; intranasal, 373; modern methods, 373–74; nonsurgical, 374; representations, 377; World War I, 372–73
Rhytidectomy, 351; antiaging facial surgery, 150–51
Roe, John Orlando, intranasal rhinoplasty, 373
Roe v. Wade, abortion legalized, 423
Roman matrimony, 276–77

Romans (ancient): antiaging, 138; birth control, 420; bodybuilding, 344–46; castration, 504; cosmetic procedures, 130; ear piercing, 95; eunuchs, 509–11; eye makeup in antiquity, 117; hair removal, 258; ideal nose, 361–62; makeup use, 143; penis as phallus, 385–86; vagina and childbirth, 532

Rothman, Barbara Katz, on medicalization of birth control, 422–23

Rubenstein, Helena: beauty industry, 145–46; history of eye makeup, 119

Saint-Hilaire, Isidore Geoffroy, founder of teratology, 219

Salkh, radical circumcision, 394

Samarran tradition (Iraq), nose piercing, 366–67

Sandow, Eugen: early years of bodybuilding, 345*f*, 346–47; fig leaf covering genitals, 389

Sanger, Margaret: advocate of family planning, 426; coined term birth control, 419, 419*f*; violation of Comstock Laws, 421–22

Sara group, Chad, lip stretching, 329–30, 331, 334–35

Scarification, 466–73, 467*f*; aesthetics, 469–70; contemporary, 471–73; folk surgery procedure, 133–35; modern practices, 472–73; multivalence, 470–71; reasons for decline in traditional societies, 471; rites of passage, 468–69; skin cutting, 464; Western society, 471–72

Scar revision, facial skin treatment, 131–32

Schizophrenia: chlorpromazine for treatment, 19; evil eye and, 111

Scrotal ladder, 234

Scrotal piercing, 234

Second Serve (Richards) 1986, autobiography of transsexuality, 244

Self-flagellation, early Christians, 5

Self-injurious behaviors (SIB), skin cutting, 464–66

Semen, **437–45**; cultural history, 437–45; fertility factor, 440–41; fluid tenacity, 444; how scientists see sperm, 439–40

Sexologists, read on clitoris, 81

Sexology magazine, transsexualism, 238

Sex reassignment surgery, 235–44; cultural representations, 243–44; debates and controversies, 241–42; definition, 235; globally, 242–43; history, 237–39; medical developments, 240–41; physiology, 239–40; popular destinations for, 243; transsexuality and, 236–37

Sexual Behavior in the Human Female (Kinsey), 523; clitoris, 81

Sexual behavior/Sexuality: blood sports, 6; breasts and, 32–33; buttocks and anus eroticization, 61–62; castration and, 503; chastity belt, 517–19; clitoris, 80–85, 522–23; designer vagina, 525; female orgasm, 522–23; fetishized feet, 199–200; footbinding and, 206; lipstick and, 326; mammary madness, 41–42; menopause effects, 403; penis envy, 396–98; pubic hair removal and, 264–65; *The Vagina Monologues*, 536–37; vaginal-tightening surgery, 537–39

Sexually transmitted diseases, circumcision effects on transmission, 393

Shadow on a Tightrope (Hannah, Schoenfielder, and Wieser), fat and feminism, 184

Shaving, facial and body hair, 259

Shoes: early history, 210–11; in fantasy and fetish, 214; high-heel, 214; history of, 208–15; medieval and renaissance Europe, 212–14; variety, 209–10, 209*f*

Sign language, hand communications, 266

Silent Spring Institute (SSI), environmental pollutants and women's health, 23

Silicone: breast enhancement, 50–51, 50*f*; cheek implants, 70; chin implants, 79; controversy on breast implants, 55–56; lip enhancement, 326–27; nose implants, 374; post-mastectomy reconstruction, 56

Skilled craftsmanship, working with ones hands, 269

Skin, **446–92**; aging process, 455–56; biology, 455; bleaching, tanning, and modern aesthetics, 463–64, 473–76; body piercing, 446–52; branding, 452–54, 472–73; class and color, 462–63; color, 457; cultural history, 454–64; cutting, 464–66; environment, 458–59; genes, 457–58; race and color, 459–62; scarification, 466–73; stretch marks, 476–77; subdermal implants, 478–79; tattoos, 479–89

Skin color map, Renato Biasutti, 460

Skinhead groups, hair removal for symbolic reasons, 257

Skin lightening, 473–76; health risks, 475; methods, 475

Skin resurfacing, antiaging treatments, 154–55

Skoog, Tord, neck lift, 351

Skull, **490–92**; trephination, 490–92

Slaves: branding, 453; skin color and, 461

The Sluggard (Leighton), depiction of male form, 76

Snakebite piercing, lips, 330

Soemmerring, Samuel Thomas, organ of the soul, 16

Some Thoughts Concerning Education (Locke), 542

Soul, concept of, 12

South Beach Diet, 174–75

Southeast Asia: ear piercing, 96; earlobe stretching, 101

Spallanzani, Lazzaro, scientists view of sperm and fertilization, 440

Spence, Jo, experience with breast cancer, 44

Sperm: normalizing, 441; research, 440; sacred, 437–39; scientists view of, 439–40. *See also* Semen

Sperm competition, 441–42; popularizing, 443–44

Sperm count, 441

Spermisists, 17th Century scientists, 380

Spermism theory of preformation, 438–39

Sperm morphology, 441

Sperm motility, 441

Spirituality, hands as a reflection of, 270

Sports, hand dexterity and, 269

Spurzheim, Johann, phrenology in the U.S. and U.K., 285
Stanford University, sex reassignment surgery, 239
Stanton, Elizabeth Cady, fashion as an individual choice, 544
St. Augustine, beginning of life, 380
Steatopygia, 63
Steroids, 348–49
Stockton, Abbye, female bodybuilding, 347
Strauss, Levi, invention of jeans, 315
Stretch marks, 476–77; prevention and treatment, 477
St. Thomas Aquinas: beginning of life, 380; importance of seminal fluid, 437–38; mind-body problem, 14
Studies in the Psychology of Sex (Ellis), 61
Subcutaneous musculoaponeurotic system (SMAS), rhytidectomy, 150–51
Subdermal implants
Subincision, radical circumcision, 395, 398–401
Suction-assisted lipectomy (SAL), 1, 192
Summa theologica II (St. Thomas Aquinas), 437–38
Sunglasses, 125
Supernumerary nipple, 358–59
Surgery: abdominoplasty, 3; bariatric, 177–78; for beauty, 130–33; body-lift, 195–96; breast reconstruction after cancer, 56; breast reduction and enlargement, 52–58; buttock reshaping, 66–68; castration, 502–7; cheekbone reshaping, 69; chin reshaping, 77; circumcision, 391; fat reduction, 191–96; folk, 133–35; labiaplasty, 311–13; LASIK, 126; limb-lengthening, 318, 321–24, 322f; lip enhancement, 326–27, 332; lip lift, 333; neck lift, 351–53; otoplasty, 102–4; pectoral implants, 74–76; reshaping of the jaw, 308–10; restoration of the hymen, 306–7; rhinoplasty, 363–64, 370–78; sex reassignment surgery, 235–44; sexual enhancement, 525; trans breast, 57; vaginal repair, 533;

vaginal-tightening, 537–39. *See also* Cosmetic surgery
Surma group, Ethiopia, lip stretching, 330, 331, 335
Surrogate mothers, India, 412f
Susan G. Koman Breast Cancer Foundation, 23
Sushruta Samhita: early otoplastic techniques, 103; rhinoplasty, 371
Suya tribe (Brazil): ear piercing, 97, 447–48; lip stretching, 335, 447–48
Symmetry, in facial beauty, 127–28
Syndactylism, 198

Tabular vault modification, 288
Tabwa (Congo), scarification, 470
Tamoxifen, breast cancer treatment, 22, 24
Tampons, 434–36
Tanning, skin darkening, 464
Tarya Neolithic people, nose piercing, 367
Tattooing, 479–89; contemporary Europe and U.S., 486–89; cultural context, 481–86; gang affiliation, 268; method of adorning hands, 267–68; penis, 389
Teeth, **493–501**; cosmetic dentistry, 493–97; cultural associations, 340–42; filing, 497–501, 498f; pain and, 342–43
Teratology (physical anomalies), 219
Testicles, **502–13**; castration, 502–7; eunuchs, 507–11
Testicular cancer, orchiectomy and orchidectomy, 502
Test-tube babies, 413–14
Thailand, kathoes (intersexual individuals), 216
Thaipusam festival, piercing and flesh-hook pulling, 448
Their Eyes Were Watching God (Hurston), straight hair in the story, 262
Theory of the Leisure Class (Veblen), corset, 541–42
Thermalift, anti-aging skin treatment, 456
Thigh, **512–13**; liposuction, 512–13
Three Essays on Sexuality (Freud), erotic qualities of the breast, 36
Three Essays on the Theory of Sexuality (Freud), 522–23

Titus, Simon David, monograph on menopause, 404
Tonga, subincision, 400
Tongue, **514–16**; legislation outlawing splitting, 515; splitting, 514–16
Toxic Links Coalition (TLC), toxin links to breast cancer, 23
Transamerica, 2005, 244
The Transfeminist Manifesto (Koyama), 242
The Transsexual Empire (Raymond), 241
Transsexualismus, term coined by Hirschfeld, 238
Transsexuality: breast surgery in, 57; cinematic representations, 244; cosmetic facial surgery, 132; cultural representations, 243–44; mandibular angle reduction surgery, 310; medicalization of, 242; reassignment surgery, 236–37; reduction malarplasty, 71
The Transsexual Phenomenon (Benjamin), 238
The Trauma of Birth (Rank), 535
Trephination, 490–92. 491f; Asia and Eastern Europe, 492; South American and Egypt, 492
Triangle piercing, 233
Trotula, vagina in, 520–21
Truth, Sojourner, on breastfeeding, 41
Tumescent liposculpture, 512–13
Tummy tuck, 1, 192
Turkey (ancient), pubic hair removal, 264

U.S. Food and Drug Administration (FDA): cellulite reduction, 166; weight loss drugs, 170
Ubangi group, Chad, lip stretching, 329–30, 331, 334–35
Ultrasonography, enhancement of vision, 113–14
Ultraviolet (UV) radiation, skin damage, 458
Umezawa, Fumio, Asian blepharoplasty, 108
Unbearable Weight (Bordo): on beauty ideals, 135; on dieting for women, 176

United Nations Department of Economic and Social Affairs, statistics on birth control use worldwide, 425–26
United States Agency for International Development (USAID), voluntary family planning, 426
University of California, Los Angeles (UCLA), sex reassignment surgery, 239
Usog, Filipino version of evil eye, 111
Uterus, **517–31**; chastity belt, 517–19. 518*f*

Vagina, **532–39**; childbirth, 532–35; cultural history, 519–26; cultural representations and practices, 523–25; designer, 525; the material mouth, 532; pelvic exercises, 525; physiology, 520; reclaiming, 525–26; tightening surgery, 537–39; venerated, 523; Western medicine and culture, 520–21
Vagina dentata, 506, 524
Vaginal orgasm, 522–23
The Vagina Monologues (Ensler), 534, 536–37, 536*f*
Vaginoplasty, 525; in sex reassignment surgery, 239–40
Valois, type of shoe in medieval Europe, 213
van Leeuwenhoek, Antoni, microscope, 439
Veiling, 292–97, 292*f*; Christianity, 295–96; Islam, 293–95; Judaism, 295; males, 296
Venesection, 4–6
Venus of Willendorf, breast representation, 34
Via Recta (Venner), first reference to obesity, 173

Vibrator, early history of, 522
Virginia Black Code, measurement of racial status, 9–10
Virginity, restoration of the hymen, 306–7
Vision: anatomy and physiology, 114; cultural significance, 112–13; ophthalmology, 114–15; technological enhancement, cultural significance of, 113–14
Vitamin D hypothesis, skin colors, 458
Vitiligo, 159–161. 474
Vogue Beauty Book, 1933, 120
Voigtländer, Johann Freidrich, monocle, 125
Voluntary motherhood, birth control in the late 19th Century, 420–21
Von Gräfe, Karl Ferdinand, modern blepharoplasty procedure, 105, 151
von Hartsoeker, Nicolass, preformation theory, 380
von Luschan scale, skin color, 459–60
von Soemmering, Samuel Thomas, reference regarding Black/African body, 1784, 327–28

Waist, **540–45**
Walker, Madame C. J. (Sarah Breedlove), beauty industry, 145–46
Wandering womb, 526–31; female physical nature, 529–30; symptoms, 527–28; treatment, 528–29
Ward, Jim, innovator in body piercing, 450
Watts, James, lobotomy in the U.S., 18–19
Weight loss: abdominoplasty after, 2; surgeries, 191–96

Weight-loss companies, 175
Western culture, ear piercing, 98
Wet-nursing, 25–27; breasts for hire, 37–38; early historical details, 34–35
What's Wrong with Addiction? (Keane), on overeaters, 171
White Teeth (Smith), hair straightening in African Americans, 262
Wigs, 254–55
Williams Obstetrics, primary medical reference, 432–33
Willis, Thomas, mind-body problem, 15–16
Winslow, Jean-Baptiste, diatribe against the corset, 542
Wolff, Caspar Friedrich, epigenesis theory, 380
The Woman Rebel, Margaret Sanger's socialist journal, 420
Women's Liberation Party, bra protest, 45, 45*f,* 48–49
World Health Organization (WHO): female genital cutting, 224, 228, 524; obesity, 169, 177; parasuicide, 466
World Professional Association for Transgender Health (WPATH), 236

Xiajiadian, ancient earrings, 94
X-rays, enhancement of vision, 113

Yakuza (Japan), subdermal implants in penis, 478
Yoruba (Benin and Nigeria), scarification, 470
Yungar (Niger-Congo), scarification, 469

Zeiss Optical Works, lens production, 124
Zone diet, 174

ABOUT THE EDITOR AND CONTRIBUTORS

Victoria Pitts-Taylor is Professor of Sociology, Queens College and the Graduate Center, City University of New York, and co-editor of the journal *Women's Studies Quarterly*. She specializes in social theory, the sociology of medicine, the sociology of the body, and is the author of *Surgery Junkies: Wellness and Pathology in Cosmetic Culture* (2007), *In the Flesh: The Cultural Politics of Body Modification* (2003), and numerous articles.

Aren Z. Aizura is a doctoral candidate in cultural studies at the University of Melbourne in Australia.

Susan E. Bell is Professor of Sociology and A. Myrick Freeman Professor of Social Sciences at Bowdoin College in Brunswick, Maine.

Erynn Masi de Casanova is a doctoral candidate in sociology at the Graduate Center, City University of New York, New York.

Paul Cheung is an honorary associate at the Centre for Values, Ethics and the Law in Medicine, University of Sydney, Australia.

Bonnie French is a doctoral student in sociology at the Graduate Center, City University of New York, New York.

Barry Gibson is Senior Lecturer in Medical Sociology in the Department of Oral Health and Development Department, School of Clinical Dentistry, University of Sheffield, United Kingdom.

Debra Gimlin is Lecturer in Sociology at the University of Aberdeen, Scotland, United Kingdom.

Michele Goodwin is the Everett Fraser Professor of Law and a professor in the schools of medicine and public health at the University of Minnesota, Minneapolis.

Angela Gosetti-Murrayjohn is an Associate Professor of Classics at the University of Mary Washington in Richmond, Virginia.

Andrew Greenberg is a doctoral candidate in sociology at the Graduate Center, City University of New York, New York.

Birgitta Haga Gripsrud is an independent scholar who has lectured at the School of Fine Art, History of Art & Cultural Studies at the University of Leeds, United Kingdom.

Martine Hackett is a program coordinator and researcher at the New York City Department of Health and Mental Hygiene.

Angelique C. Harris is an assistant professor in the Department of Sociology at California State University, Fullerton.

Margaret Howard is an independent scholar and English teacher in Wurtzburg, Germany.

Kay Inckle is IRCHSS Post-Doctoral Research Fellow in the School of Social Work and Social Policy, Trinity College, Dublin, Ireland.

Mike Jolley is a doctoral candidate in sociology at the City University of New York Graduate Center and an adjunct lecturer in sociology at Hunter College, New York.

Meredith Jones is a lecturer at the Institute for Interactive Media and Learning at the University of Technology, Sydney, Australia.

Barbara Katz Rothman is Professor of Sociology at the City University of New York.

Matthew Lodder is a doctoral candidate in the history of art and architecture at the University of Reading in Reading, England.

Lauren Jade Martin is a doctoral candidate in sociology at the Graduate Center of the City University of New York.

Laura Mauldin is a doctoral candidate in sociology at the Graduate Center, City University of New York, New York.

Jay Mechling is Professor of American Studies at the University of California, Davis.

Zoë C. Meleo-Erwin is a doctoral candidate in sociology at the Graduate Center, City University of New York, New York.

Lisa Jean Moore is an associate professor of sociology and coordinator of gender studies at Purchase College, State University of New York, Purchase, New York.

Samantha Murray is a postdoctoral research fellow in the Department of Critical and Cultural Studies and Somatechnics Research Centre, Macquarie University, Sydney, Australia.

Travis Nygard is a visiting instructor at Gustavus Adolphus College, St. Peter, Minnesota, and a doctoral candidate at the University of Pittsburgh, Pittsburgh, Pennsylvania.

Gretchen Riordan is a doctoral candidate in cultural studies at Macquarie University, Sydney, Australia.

Peter G. Robinson is professor and head of the Department of Oral Health and Development, School of Clinical Dentistry, University of Sheffield, United Kingdom.

Katarina Rost is an independent midwife in Nuremberg, Germany.

Alena J. Singleton is a doctoral candidate in sociology at Rutgers University, New Brunswick, New Jersey.

Emily Laurel Smith is a doctoral student in American studies at the University of Minnesota, Twin Cities, Minnesota.

Alec Sonsteby is an Instruction and Reference Librarian at Concordia College, Moorhead, Minnesota.

Alyson Spurgas is a doctoral candidate in sociology at the Graduate Center, City University of New York, New York.

Nikki Sullivan is associate professor of critical and cultural studies at Macquarie University, Sydney, Australia.

Karen Throsby is an associate professor in the Department of Sociology at the University of Warwick, United Kingdom.

Kara Van Cleaf is a doctoral candidate in sociology at the Graduate Center, City University of New York, New York.

Sean Weiss is a doctoral candidate in art history at the Graduate Center, City University of New York, New York.

Alana Welch is an independent scholar and studied sociology at the Graduate Center, City University of New York.

Jaime Wright is a doctoral student in ethics and social theory at the Graduate Theological Union, Berkeley, California.